Lecture Notes in Artificial Intelli

Subseries of Lecture Notes in Computer Science
Edited by J. G. Carbonell and J. Siekmann

Lecture Notes in Computer Science

Edited by G. Goos, J. Hartmanis and J. van Leeuwen

Springer
Berlin
Heidelberg
New York
Barcelona
Budapest
Hong Kong
London
Milan
Paris
Singapore
Tokyo

Andreas Birk   John Demiris (Eds.)

Learning Robots

6th European Workshop, EWLR-6
Brighton, England, August 1-2, 1997
Proceedings

 Springer

Series Editors

Jaime G. Carbonell, Carnegie Mellon University, Pittsburgh, PA, USA
Jörg Siekmann, University of Saarland, Saarbrücken, Germany

Volume Editors

Andreas Birk
Vrije Universiteit Brussel, Artificial Intelligence Laboratory
Pleinlaan 2, B-1050 Brussels, Belgium
E-mail: cyrano@arti.vub.ac.be

John Demiris
Department of Artificial Intelligence, University of Edinburgh
5 Forrest Hill, Edinburgh EH1 2QL, Scotland, UK
E-mail: johnde@dai.ed.ac.uk

Cataloging-in-Publication Data applied for

Die Deutsche Bibliothek - CIP-Einheitsaufnahme

Learning robots : 6th European workshop ; proceedings / EWLR '96, Brighton,
England, August 1 - 2, 1996. Andreas Birk ; John Demiris (ed.). - Berlin ;
Heidelberg ; New York ; Barcelona ; Hong Kong ; London ; Milan ; Paris ;
Singapore ; Tokyo : Springer, 1999
 (Lecture notes in computer science ; 1545: Lecture notes in artificial
intelligence)
ISBN 3-540-65480-1

CR Subject Classification (1998): I.2.9, I.2.11, I.2.6, I.6

ISBN 3-540-65480-1 Springer-Verlag Berlin Heidelberg New York

Typesetting: Camera ready by author
SPIN 10693075   06/3142 – 5 4 3 2 1 0   Printed on acid-free paper

Preface

Robot learning is a broad and interdisciplinary area. This holds with regard to the basic interests and the scientific background of the researchers involved, as well as with regard to the techniques and approaches used. The interests that motivate the researchers in this field range from fundamental research issues, such as how to constructively understand intelligence, to purely application oriented work, such as the exploitation of learning techniques for industrial robotics. Given this broad scope of interests, it is not surprising that, although AI and robotics are usually the core of the robot learning field, disciplines like cognitive science, mathematics, social sciences, neuroscience, biology, and electrical engineering have also begun to play a role in it. In this way, its interdisciplinary character is more than a mere fashion, and leads to a productive exchange of ideas.

One of the aims of EWLR-6 was to foster this exchange of ideas and to further boost contacts between the different scientific areas involved in learning robots. EWLR is, traditionally, a *"European* Workshop on Learning Robots". Nevertheless, the organizers of EWLR-6 decided to open up the workshop to non-European research as well, and included in the program committee well-known non-European researchers. This strategy proved to be successful since there was a strong participation in the workshop from researchers outside Europe, especially from Japan, which provided new ideas and lead to new contacts.

EWLR-6 was held in conjunction with ECAL'97, the "European Conference on Artificial Life". EWLR-6 was an independent event, but it took place at the same site, as a two-day workshop starting immediately after the main conference. This allowed an interaction between Robot Learning and Artificial Life, bringing out some of the common methodological approaches between the two fields, mainly the investigation of fundamental research issues through constructive approaches. Of course, Alife is only one of the directions which EWLR intends to investigate and provide with a forum. There are many more approaches in the field of learning robots, and the range of papers included here in the proceedings of EWLR-6 reflects this broad range of views.

The proceedings start with a chapter related to an invited talk, where Luc Steels presents a view of learning robots in the context of research on the evolution of language. The related chapter by Belpaeme, Steels, and Van Looveren describes the learning of visual categories by a real robot, thus presenting a possible way to bridge the gap between dynamic, real-world sensing and the symbolic level of language acquisition.

The following three chapters of the book deal with reinforcement learning. Murao and Kitamura in their chapter describe an efficient method of constructing the state space by segmenting a sub-region of the sensor space or recombining existing states, starting from a single state covering the entire sensor space.

This method allows the Q-learning algorithm to accomplish their task in continuous sensor space. Kalmar, Szepesvari, and Lorinez in their chapter propose using high-level modules in order to transform the task into a completely observable one, and compare their method with several other RL variants on a real-robot task. Finally, Faihe and Müller introduce a methodology for designing reinforcement-based control architectures.

Uchibe, Asada, and Hosoda also deal with the state-space construction issue with an emphasis on a vision-based approach in a multi-agent environment, and apply their method to soccer playing robots. Of course, other agents in the environment do not always simply add an extra level of complexity, but can also be utilized to help a robot to learn how to achieve new tasks. The next two chapters deal with how to achieve that through imitation. Billard and Hayes propose using imitation to teach a vocabulary to a learner robot who is following the teacher robot, using a recurrent associative memory as the underlying architecture, while Grossmann and Poli present an approach that combines learning by imitation with reinforcement learning and incremental hierarchical development.

Robots, in order to be useful, need not only to be able to learn but also to be self-sufficient, and Birk in his chapter deals with the interplay between learning and self-sufficiency. They also need the ability to learn their environment. Vogt attempts to ground an adaptive language that can be used to describe objects in the robot's environment, while Chagas and Hallam use case-based reasoning to capture regularities in its environment and reinforcement learning to gradually improve the acquired knowledge and the robot's performance in a navigation task.

Finally, the interplay between evolutionary methods and learning are the subject of the two last papers. Lee, Hallam, and Lund, in their chapter, describe how complex robot behaviors can be learned using an evolutionary approach coupled with task decomposition, while Keymeulen et al. attempt a similar goal but at a completely different level, by evolving hardware at the gate level. These last two papers are also a good example of what can be achieved by the interaction between the fields of Artificial Life and Robotics.

Andreas Birk and John Demiris
Brussels and Edinburgh, 1998

Acknowledgments

The workshop organizers are especially indebted to the members of the program committee of EWRL-6, who spend quite some time and effort for the reviews and the selection process.

Program Committee

- Minoru Asada   (Osaka University, Japan)
- Luc Berthouze   (Electrotechnical Laboratory, Japan)
- Hendrik Van Brussel   (Katholieke Univ Leuven, Belgium)
- Thomas Christaller   (GMD FIT, Germany)
- Edwin DeJong   (Vrije Univ Brussel, Belgium)
- Rüdiger Dillmann   (Univ of Karlsruhe, Germany)
- Marco Dorigo   (Univ Libre de Bruxelles, Belgium)
- Attilo Giordana   (Univ of Turin, Italy)
- John Hallam   (Univ of Edinburgh, Scotland, UK)
- Gillian Hayes   (Univ of Edinburgh, Scotland, UK)
- Yasuo Kuniyoshi   (ETL, Japan / MIT, USA)
- Ramon Lopez de Mantaras   (IIIA, Barcelona, Spain)
- Rolf Pfeifer   (Univ of Zurich, Switzerland)
- Fabio Roli   (Univ of Cagliari, Italy)
- Jun Tani   (Sony CS Laboratory, Japan)
- Sebastian Thrun   (Carnegie Mellon Univ, USA)
- Walter Van de Velde   (Vrije Univ Brussel, Belgium)
- Jeremy Wyatt   (Univ of Birmingham, UK)

Table of Contents

The Construction and Acquisition of Visual Categories

Tony Belpaeme[1], Luc Steels[1,2], and Joris Van Looveren[1]

[1] Vrije Universiteit Brussel
Artificial Intelligence Laboratory
Pleinlaan 2, 1050 Brussels, Belgium
steels@arti.vub.ac.be
http://arti.vub.ac.be
[2] Sony Computer Science Laboratory Paris

Abstract. The paper proposes a selectionist architecture for the construction and acquisition of visual categories by a robotic agent. The components of this architecture are described in some detail and results from experiments on physical robots are reported.

Keywords: origins of language, self-organization, distributed agents, open systems, vision.

1 Introduction

A cognitive agent must have a way to map symbolic categories onto its experiential, perceptually acquired data streams. For example, in a task requiring language communication or symbolic planning, the data gathered from visual input through a camera must somehow be categorised and interpreted. The problem of bridging the gap between the subsymbolic world of continuous dynamics and real world sensing and effecting and the symbolic world of categories and language, has turned out to be very difficult. No successful general purpose vision system has emerged that is capable to perform the mapping task in real-time. Part of this is due to the complexity of the task, requiring vast amounts of computation power and memory. It is also due to the enormous richness of real-world data and the difficulty to extract regularity.

So far two approaches dominate the field. The first approach, pioneered and exemplified by the work of David Marr [5] consists in programming explicitly complex feature detectors which operate on successive stages of information processing. The end result is a symbolic 3-d model of the scene. Although a large amount of practically usable image processing components have resulted from this approach, the ultimate goal of a general purpose vision system that could interface with a symbolic knowledge representation component has not materialised. Some researchers who have at one point been fully involved with this approach, notably Brooks [14], have decided to give up altogether and insist on staying at the subsymbolic level, at least for the control of behaviors. Others such as Ullman [15] have introduced more focus by limiting the scope of vision to special-purpose detection handled by so called visual routines.

A second approach is associated with neural network modeling. It relies on learning mechanisms to acquire the feature detectors by continuous exposure to a series of examples and counter examples. This approach has also a long tradition, starting with the perceptron, and advancing with the discovery of the back-propagation learning algorithm, Kohonen's self-organising feature maps, etc. In general, the neural network approach has been able to come up with solutions that are robust against the natural variation occuring in real-world data. But there are two difficulties. The supervised learning systems rely on an outside intelligence to come up with the initial series of examples. Such a setup is not possible for autonomous robotic agents which may find themselves in environments that are unknown to humans or perceivable through sensory modalities (such as infrared) to which humans have no access. The unsupervised learning systems on the other hand are only constrained by the regularities in the data themselves. They are inductive mechanisms that cannot be steered to develop categories under the influence of certain tasks.

This paper reports on experiments which explore an alternative to the approaches briefly introduced above. Our first hypothesis is that the vision module should be made responsible for less. We make a distinction between three components:

1. The *vision component* which is responsible for segmenting and data collection.
2. The *categorisation component* which performs the transition from real world data to symbolic descriptions.
3. The *user component* which is another process that needs the symbolic descriptions in one way or another. This could be a language task or a planning task.

Depending on the task (i.e. the user component) the categorisation may be different. So no general purpose visual categorisation is sought for, on the contrary. The categories should be adapted to the task.

Our second hypothesis is that each component is constructed and adapted through a selectionist process. A selectionist process contains on the one hand a generator of diversity which is not driven by the task at hand, and on the other hand a process maintaining or eliminating variations based on their performance, i.e. how well they do with respect to a set of selectionist pressures provided by the context or users of the result. Evolution of species by natural selection, as proposed by Darwin, or reenforcement of natural variation in the immune system, as proposed by Jerne and Edelman, are two examples of selectionist processes. But the idea can be applied to any system that can be made to exhibit spontaneous variation and subsequent selection.

This paper focuses only on how the categorisation component could self-organise. We assume that category buildup proceeds by the spontaneous creation of new distinctions when the user component provides feedback that the existing set of categories are insufficient. Whether the new distinction is adequate will depend on later evaluations. The selectionist pressure in other words comes from the user component. The same scheme could be applied to the vision component. In particular, the vision component could internally have a generator of diversity that spontaneously creates new data if the categorisation module provides feedback that the existing data is not sufficient to construct adequate categorisations. The categorisation module acts in this case as the source of selectionist pressure.

To make the paper concrete, we assume that the user component is a language system that is itself also selectionist. The language system lexicalises categories or combinations of categories called feature structures. Selectionist pressure now comes from whether a particular lexicalisation gains acceptance in the rest of the community. In this paper, the "talking heads" experiment is taken as a source of examples. In this experiments two robotic heads which can track moving objects based on visual input, watch a static or dynamic scene. The robots must develop from scratch visual categories which are adequate for describing what they see. The language itself also develops from scratch and without human intervention.

Other papers have provided details on the language component and how this component interacts with a (selectionist) categorisation module (see e.g. [8] or [10]). This paper focuses on the interaction between the vision component and the categorisation component. The vision component itself is assumed to provide a fixed set of data. The present research has first been validated with software simulations and then ported to real world physical robots (see also [13]).

The rest of the paper is in four parts. The next part describes the robotic set-up used in the present experiments. The third part describes the vision component as it is currently implemented. The fourth part describes the categorisation component. Then the results of some experiments are presented in some detail.

2 The Talking Heads Experiment

A *talking head* consists of a black and white camera mounted on a pan-tilt unit, electronics to perform low-level signal processing and actuator control, and two PCs performing visual processing and symbol processing respectively. The experiment uses two heads. Both stand next to each other, observing the same scene. This scene contains several objects in various shapes and sizes. The scene can even contain moving objects, such as a robot driving around and pushing other objects.

The visual processing filters the incoming image from the head. In the next step, some visual modules are unleashed on the filtered image. A module is specifically for a certain feature of the image, e.g. a specific color. In the current implementation three modules are used: one module is active for motion in the image, one for detecting patches of a light intensity and one for detecting patches with a dark intensity. The module detecting motion in the image is also connected to the control of the pan-tilt motors of the head. This results in the heads focussing on motion in the scene.

Fig. 1. The eye of a "talking head" with associated electronics for low level signal processing and real-time actuator control needed for tracking.

Each visual module singles out patches in the image which are aggregated to segments withing a rectangular bounding box. Overlapping segments are joined and segments that are too small for further analysis neglected. In a next step, data for each patch (such as average intensity, position in the image, ...) is calculated. All these patches and their features are continuously being logged, along with the first derivative in time of each feature (as to detect changes over time) and the time at which a patch appeared or disappeared in the image.

After a few moments of observing the scene, the heads agree to communicate. At this moment, all information on the patches in the image logged up until now is passed on to the categorisation and lexicon components. When this is finished, the heads continue observing the scene.

3 The visual processing

The visual perception is inspired by active vision (for an overview see [1] or [2]). The early visual processing contains bottom-up, uniform and unarticulated processing of the raw image; such as filtering or transforming the image in a more useful representation. This representation is then fed to the visual modules responsible for producing the data used in the language formation. The three modules used are now described in some detail.

The motion module detects motion in the image. This is done by subtracting two subsequent image frames. $f(x, y, t_j)$ is an image taken at time j, $motion(x, y)$ is an array which will contain a 1 if pixel (x, y) moved, or a 0 if not. The threshold $\vartheta > 0$ is used to filter noise.

$$motion(x, y) = \begin{cases} 1 \text{ if } |f(x, y, t_{i+1}) - f(x, y, t_i)| > \vartheta \\ 0 \text{ otherwise} \end{cases}$$

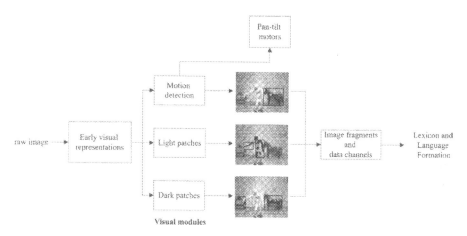

Fig. 2. schematic overview of the visual processing for one talking head.

A bounding box is placed around a consistent set of moving pixels.

The centroid of all moving pixels is also calculated and passed on to the control of the pan and tilt motors of the head. When the centroid tends to move outside the image, the motors of the head are repositioned in order to get the centroid of motion back in the middle of the image. The head will always focus on regions with lots of movement, e.g. a robot driving around or a person walking by. The camera does not move as long as the centroid is near the middle of the image, thus resembling saccadic eye movements observed in mammals. Hence we call this *saccadic tracking.*

The light patches module returns all segments in the image of size $w \times h$ having a high intensity. First the average intensity $\bar{I}$ of all pixels and the standard deviation δ of the intensity are calculated.

$$\bar{I} = E(f(x, y, t_i)) \quad \text{with} \quad E(x) = \frac{\sum x}{w.h}$$

$$\delta = \sqrt{E(f^2(x, y, t_i)) + (E(f(x, y, t_i)))^2}$$

Next, every pixel is checked for being lighter than $\bar{I} + \delta$. $light(x, y)$ is an array containing 1 at position (x, y) if the pixel is light, and 0 if not.

$$light(x, y) = \begin{cases} 1 \text{ if } f(x, y, t_i) > \bar{I} + \delta \\ 0 \text{ otherwise} \end{cases}$$

The dark patches module is the same as the previous module, but now pixels with an intensity lower than $\bar{I} - \delta$ are considered.

$$dark(x, y) = \begin{cases} 1 \text{ if } f(x, y, t_i) < \bar{I} - \delta \\ 0 \text{ otherwise} \end{cases}$$

Image segments are groups of consistent pixels. The way image segments are made is identical for each visual module. Here the algorithm, which resembles region growing from traditional computer vision, is explaned for the light patches module.

1. First, a non-zero entry (a, b) is picked from the array $light(x, y)$. Add (a, b) to a new, empty image segment.
2. For each pixel in the image segment: if a neighboring pixel is non-zero, add it to the image segment. Delete the entry in array $light(x, y)$.
3. Redo the previous step, until all neighboring pixels are zero.
4. Calculate a bounding box around this image segment. If an image element has less than β pixels (an arbitrary threshold) then it is not used in further processing.
5. Take a new non-zero entry, and start from step 1. Stop if no non-zero entries are left in $light(x, y)$.

It should be noted that no object recognition is involved; the visual processing does not recognize or select objects. Anything in the scene, even the floor or the background, can and will be analyzed if it triggers a visual module.

An image segment is a region in the image where every pixel shares a common property. Now, for every image segment, some *sensory channels* are calculated. Sensory channels are real-valued properties of a restricted region in an image. The following sensory channels are used:

Vertical angle of the image segment, relative to the observer (i.e. the talking head).

Horizontal angle of the image segment.

Average intensity of the image segment.

Area of the image segment. This correlates typically with the distance to the observer. The further an image segment is from the observer, the smaller its area will be.

Visibility is a measure for how close an image segment is to the focus of attention: the closer the segment is to the FOA, the higher the visibility will be. The FOA is determined by the centroid of motion, calculated by the motion module.

Fill ratio is the ratio of the pixels marked as interesting by the visual module and the size of the rectangular bounding box.

In figure 3 the output of the visual processing is shown. The scene is static and contains a house, a wooden puppet and a toy horse. Bounding boxes are placed around interesting image segments. The transcript (see table 1) of the data produced by the visual processing shows each image segment, with the coordinates of its bounding box, the start and stop time in milliseconds at which it was visible and the data calculated for each segment. Next to each data value d, $\delta d/\delta t$ is given; since the scene is static, all $\delta d/\delta t = 0$.

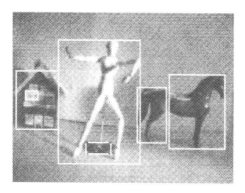

Fig. 3. A screenshot of the output of the visual processing.

4 Ontology creation through discrimination games

We now turn to the categorisation module which starts from the output of the vision module.

Let there be a set of segments $\mathcal{O} = \{o_1, ..., o_m\}$ and a set of sensory channels $S = \{\sigma_1, ..., \sigma_n\}$, being real-valued partial functions over $\mathcal{O}$. Each function σ_j defines a value $0.0 \leq \sigma_j(o_i) \leq 1.0$ for each segment o_i.

An agent a has a set of feature detectors $D_a = \{d_{a,1}, ..., d_{a,m}\}$. A *feature detector* $d_{a,k} = \langle p_{a,k}, V_{a,k}, \phi_{a,k}, \sigma_j \rangle$ has an attribute name $p_{a,k}$, a set of possible values $V_{a,k}$, a partial function $\phi_{a,k}$, and a sensory channel σ_j. The result of applying a feature detector $d_{a,k}$ to an object o_i is a feature written as a pair $(p_{a,k}\ v)$ where p is the attribute name and $v = \phi_{a,k}(\sigma_j(o_i)) \in V_{a,k}$ the value.

The *feature set* of a for o_i is defined as $F_{a,o_i} = \{(p_{a,k}\ v) \mid d_{a,k} \in D_a, d_{a,k} = \langle p_{a,k}, V_{a,k}, \phi_{a,k}, \sigma_j \rangle, v = \phi_{a,k}(\sigma_j(o_i))\}$. Two features $(a_1\ v_1)$, $(a_2\ v_2)$ are *distinctive* iff $a_1 = a_2$ and $v_1 \neq v_2$. A distinctive feature set D_{a,o_t}^C is a set of features distinguishing an segment o_t from a set of other segments C. $D_{a,o_t}^C = \{f \mid f = (p\ v) \in F_{a,o_t}$ and $\forall o_c \in C$ either $/\exists f' = (p'\ v') \in F_{a,o_c}$ with $p = p'$ or $\exists f' \in F_{a,o_c}$ with f and f' distinctive$\}$. Clearly there can be several distinctive feature sets for the same o_t and C, or none.

A discrimination game $d = <a, o_t, C>$ involves an agent a, a topic $o_t \in C \subseteq \mathcal{O}$. C is called the context. The outcome of the game is twofold. Either a distinctive feature set could be found, $D_{a,o_t}^C \neq \emptyset$, and the game ends in success, or no such feature set could be found, $D_{a,o_t}^C = \emptyset$, and the game ends in failure.

As part of each game the repertoire of meanings is adjusted in the following way by the agent:

1. $D_{a,o_t}^C = \emptyset$, i.e. the game is unsuccessful. This implies that there are not enough distinctions and therefore $\exists o_c \in C$, $F_{a,o_t} \subseteq F_{a,o_c}$. There are two ways to remedy the situation:

Table 1. Sensory channels corresponding to figure 3. Each line shows the coordinates (left, top, right, bottom) of the bounding box, the start and stop times in milliseconds when the box was seen and the values for the different sensory channels together with their relative change during the time interval. The data channels respectively are: the horizontal and vertical angle, the visibility, the area, the intensity and the fill-ratio. Since this is a still image, all first derivatives in time are equal to zero.

```
((33 19 92 106)    50846 54138
  (0.465441 0.0) (0.503695 0.0) (0.515432 -0.0) (0.852174 -0.0) (0.448984 0.0) (0.388662 0.0))
((9 55 18 61)      50846 54138
  (0.399865 -0.0) (0.493691 0.0) (0.553222 0.0) (0.046518 -0.0) (0.866551 -0.0)(0.529951 0.0))
((2 41 33 82)      50846 54138
  (0.407428 0.0) (0.497312 0.0) (0.547630 -0.0) (0.188681 0.0) (0.284191 0.0) (0.332916 0.0))
((113 43 153 95)   50846 54138
  (0.570797 0.0) (0.522586 0.0) (0.453309 0.0) (0.339950 0.0) (0.145960 0.0) (0.268578 0.0))
((90 52 111 91)    50846 54138
  (0.531648 0.0) (0.528035 -0.0) (0.470159 0.0) (0.138996 0.0) (0.342711 0.0) (0.408776 0.0))
((53 91 71 99)     50846 54138
  (0.473048 0.0) (0.592732 0.0) (0.467110 -0.0) (0.025958 0.0) (0.270678 0.0) (0.542513 0.0))
((53 91 71 99)     50846 54138
  (0.470968 0.0) (0.595915 0.0) (0.466558 0.0) (0.024000 0.0) (0.337255 0.0) (0.569444 0.0))
```

(a) If there are still sensory channels for which there are no feature detectors, a new feature detector may be constructed. This option is preferred.

(b) Otherwise, an existing attribute may be refined by creating a new feature detector that further segments the region covered by one of the existing attributes.

2. $D_{a,o_t}^C \neq \emptyset$. In case there is more than one possibility, feature sets are ordered based on preference criteria. The 'best' feature set is chosen and used as outcome of the discrimination game. The record of use of the features which form part of the chosen set is augmented. The criteria are as follows:

(a) The smallest set is preferred. Thus the least number of features are used.

(b) In case of equal size, it is the set in which the features imply the smallest number of segmentations. Thus the most abstract features are chosen.

(c) In case of equal depth of segmentation, it is the set of which the features have been used the most. This ensures that a minimal set of features develops.

The whole system is selectionist. Failure to discriminate creates pressure to create new feature detectors. However the new feature detector is not guaranteed to do the job. It will be tried later and only thrive in the population of feature detectors if it is indeed successful in performing discriminations.

As mentioned earlier, the categorisation component is next coupled to a language component, which in its simplest form constructs a lexicon. The language component provides feedback about which categories have lead to successful language games, and consequently which categories made sense in conversations with other agents. More details about this language component can be found in [10].

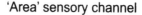
'Area' sensory channel

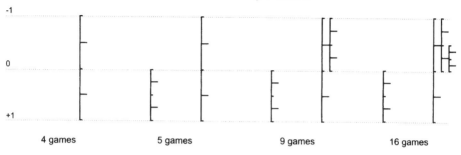

Fig. 4. An example of consecutive refinements made for the 'area' category in one of the agents. The snapshots have been taken after resp. 4, 5, 9 and 16 language games. The left axis ranges from 0 to 1 and represents the average area over a short period of time. The right axis ranges from -1 to 1 and represents the time derivative of the average area during the same period. Note that this examples is not taken from figure 3.

5 Results

Here are some examples of interactions. In the first one the speaker fails to conceptualise the scene and creates a new category by dividing the sensory channel called fill-ratio into two segments associated with the values v-81 and v-82.

```
0. Speaker: Head-16. Hearer: Head-17. Topic: o4.
   Repair Head-16:
     Extend categories: FILL-RATIO [-1.0 1.0]: v-81 v-82
   ? ? => ? ? [failure]
```

In the next game, there is another failure and a new distinction is created now on the visibility channel:

```
1. Speaker: Head-16. Hearer: Head-17.
   Repair Head-16:
     Extend categories: VISIBILITY [-1.0 1.0]: v-83 v-84
   ? ? => ? ? [failure]
```

In game 4, a set of distinctive features has been found but there is no word yet. The speaker creates a new word:

```
4. Speaker: Head-16. Hearer: Head-17. Topic: o12.
   Repair Head-16:
       Extend word repertoire: "(d u)"
       Extend lexicon: ((visibility v-88)) (d u)
   ((visibility v-88)) (d u) => ? ? [failure]
```

In the following game, the speaker is able to find a distinctive feature set and a word, but the hearer is missing the required distinctions:

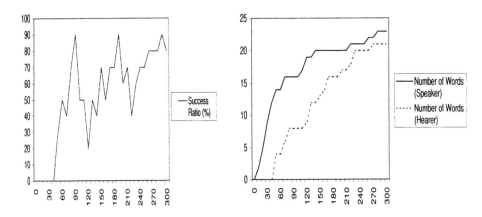

Fig. 5. Graph showing the increase in average communicative success (left) as well as the increase in the number of words in the vocabularies of two robotic heads (right).

```
6. Speaker: Head-16. Hearer: Head-17. Topic: o10.
  Repair Head-17:
    Extend categories: VISIBILITY [-1.0 1.0]: v-91 v-92
  ((visibility v-88)) (d u) => (d u) ? [failure]
```

The first successful game happens after 47 games:

```
47. Speaker: Head-16. Hearer: Head-17. Topic: o25.
  ((visibility v-109)(area v-108)(fill-ratio v-81)) (k i)
    => (k i) (fill-ratio v-125)(intensity v-134)(area v-132))
  [success]
```

A snapshot of the lexicon of one agent is as follows:

meaning	form
((visibility v-88))	(d u)
((fill-ratio v-82))	(t e)
((fill-ratio v-81)(area v-86)(visibility v-87))	(l e)
((intensity v-90))	(n a)
((fill-ratio v-81)(area v-86)(intensity v-89))	(p u)
((intensity v-89))	(m i)
((fill-ratio v-81)(area v-108) (visibility v-109))	(k i)

Figure 5 shows the increased success in communication as the agents continue to build up a shared lexicon and the increase in complexity of the lexicons.

Although the physical embodiment of the Talking Heads experiment is quite different from the mobile robots, we see the same phenomena: steady increase and adaptation of a perceptually grounded ontology, and progressive build up and self-organised coherence of a shared lexicon. The Talking Heads experiment is somewhat easier because visual perception provides a richer source for building an ontology and the communication and perceptual conditions are more stable.

6 Conclusions

We consider the experiments so far successful, in the sense that a set of visual categories emerges that are adequate for the language games that the agents play. The categories keep expanding when new segments enter in the environment.

Many extensions and variants need to be investigated further. For the moment only three visual modules and six sensory channels per module are used, this is mainly due to the limitations the black and white cameras pose. Color cameras would produce much more data, which could be used to have more visual modules and sensory channels; providing the lexicon and language formation with extended and richer data.

Second, the visual perception does not receive feedback from the lexicon or language formation. In future work, the higher cognitive processing should return information concerning usefulness of visual modules and sensory channels. This 'selectionistic pressure' could then be used to adjust the way the visual processing is done; e.g. much used visual modules could gain importance, while less used modules would die away.

Finally, rather than a rigid binary discrimination tree, it is conceivable to use a prototype approach in which the outcome of categorisation is based on comparison with a "prototype" point falling inside the possible range of values of a sensory channel. It is still possible to have a hierarchy of more or less general prototypes. Another variant would be to change the task, for example rather than finding a distinctive feature set which identifies what is different between the segments, one could focus on finding what is common between the segments. Such classification games will lead to other visual categories. It is also possible to couple the categorisation component to other user components, for example, a module that does behavioral control based on symbolic descriptions of the scene rather than continuously coupled dynamics. These many variants form the subject of current research and experimentation.

7 Acknowledgements

The design and software simulations of the discrimination games for category formation and the language games for lexicon formation have been developed at the Sony Computer Science Laboratory in Paris by Luc Steels. Several robot builders at the VUB AI Lab (soft and hardware) have contributed to the grounding experiments, including Andreas Birk, Tony Belpaeme, Peter Stuer and Dany Vereertbrugghen. The vision algorithms and the robotic heads were designed and implemented by Tony Belpaeme. Joris Van Looveren has ported the language and discrimination games to the talking heads experimental setup. Tony Belpaeme is a Research Assistant of the Fund for Scientific Research - Flanders (Belgium) (F.W.O.)

References

1. Ballard, D. (1991) Animate Vision. *Artificial Intelligence*, **48** (1991) 57-86.
2. Blake, A. and Yuille, A., editors (1992) *Active Vision*, The MIT Press, MA.
3. Edelman, G.M. (1987) *Neural Darwinism: The Theory of Neuronal Group Selection*. New York: Basic Books.
4. Kohonen, T. (1988) *Self-Organization and Associative Memory*. Springer Series in Information Sciences. Vol 8. Springer Verlag, Berlin.
5. Marr, D. (1982) *Vision* A Computational Investigation into the Human Representation and Processing of Visual Information. W.H. Freeman and Co, San Francisco.
6. Steels, L. (1994) The Artificial Life Roots of Artificial Intelligence. *Artificial Life Journal* 1(1), pp. 89-125.
7. Steels, L. (1996a) Emergent Adaptive Lexicons. In: Maes, P. (ed.) (1996) *From Animals to Animats 4: Proceedings of the Fourth International Conference On Simulation of Adaptive Behavior*, The MIT Press, Cambridge Ma.
8. Steels, L. (1996b) Perceptually grounded meaning creation. In: Tokoro, M. (ed.) (1996b) *Proceedings of the International Conference on Multi-Agent Systems*. The MIT Press, Cambridge Ma.
9. Steels, L. (1996c) Self-organising vocabularies. In Langton, C. (ed.) *Proceedings of Artificial Life V*. Nara, 1996. The MIT Press, Cambridge Ma.
10. Steels, L. (1997a) Constructing and Sharing Perceptual Distinctions. In van Someren, M. and G. Widmer (eds.) *Proceedings of the European Conference on Machine Learning*, Springer-Verlag, Berlin, 1997.
11. Steels, L. (1997b) The origins of syntax in visually grounded robotic agents. In: Pollack, M. (ed.) *Proceedings of the 10th IJCAI, Nagoya* AAAI Press, Menlo-Park Ca. 1997. p. 1632-1641.
12. Steels, L. (1997c) The synthetic modeling of language origins. *Evolution of Communication*, 1(1):1-35. Walter Benjamins, Amsterdam.
13. Steels, L. and P. Vogt (1997) Grounding language games in robotic agents. In Harvey, I. et.al. (eds.) *Proceedings of ECAL 97*, Brighton UK, July 1997. The MIT Press, Cambridge Ma., 1997.
14. Steels, L. and R. Brooks (eds.) (1995) Building Situated Embodied Agents. The Alife route to AI. Lawrence Erlbaum Ass. New Haven.
15. Ullman, S. (1987) Visual Routines. In Fischler, M. and Firschein, O. (eds.) *Readings in Computer Vision*, Morgan Kaufmann Publ., Ca. 1987. p. 298-328.

Q-Learning with Adaptive State Space Construction

Hajime Murao and Shinzo Kitamura

Faculty of Engineering, Kobe University, 1-1 Rokkodai, Nada, Kobe 6578501,
JAPAN, murao@kobe-u.ac.jp / kitamura@kobe-u.ac.jp

Abstract. In this paper, we propose Q-learning with adaptive state space construction. This provides an efficient method to construct the state space suitable for Q-learning to accomplish the task in continuous sensor space. In the proposed algorithm, a robot starts with single state covering whole sensor space. A new state is generated incrementally by segmenting a sub-region of the sensor space or combining the existing states. The criterion for incremental segmentation and combination is derived from Q-learning algorithm. Simulation results show that the proposed algortithm is able to construct the sensor space effectively to accomplish the task. The resulting state space reveals the sensor space in a Voronoi tessellation.

1 Introduction

Reinforcement learning is an efficient method to acquire adaptive behavior of a robot with little or no *a priori* knowledge of an environment where the robot will work. However, there is a problem in applying reinforcement learning to tasks in the real world, *i.e.* how to construct state spaces suitable for the reinforcement learning. A state space is usually designed by segmenting a continuous sensor space using human intuitions. Such state space is not always appropriate to accomplish a task. Coarse segmentation will cause so-called "perceptual aliasing problem" [1] [2] [3] by which a robot cannot discriminate states important to accomplish a task. Fine segmentation to avoid the perceptual aliasing problem will produce too many states to manage with computational resources such as CPU time and memory. It might be a rather reasonable solution to this problem applying a robot not with a state space designed by human but with a method to construct state space using information of its environment.

There were several approaches to construct a state space adaptively using sensor vectors. Dubrawski and Reignier [5] proposed a system employing a Fuzzy-ART neural network, for which each output neuron is defined by a representative sensor vector. When the Euclidean distance between an incoming sensor vector and its nearest representative sensor vector exceeds a given threshold value, a part of the sensor space represeted by the incoming sensor vector is subdivided into a new region by adding a new output neuron. This criterion will result in a Voronoi tessellation of a sensor space. Takahashi, Asada and Hosoda [6] proposed a method using a linear model of gradients of sensor vectors. In which, a state is defined as a set of sensor vectors with a same gradient. Their approaches made a sensor space segmentation efficiently to accomplish a task in the real world. However, we think it is rather important for accomplishing a task to segment a sensor space according to the reinforcement signals (rewards and/or penalties).

Ishiguro, Sato and Ishida [7] proposed a sensor space segmentation using rewards. Their method consists of two-steps; (1) a robot runs with the Q-learning and collects the rewards. (2) a sensor space is divided into two area according to an amount of the rewards. Fine segmentation of the sensor space is acquired by iterating this sequence. This is however not an on-line method in the strict sense. Therefore, it takes a long time to acquire a state space to accomplish a task. Kröse and Dam [4] proposed an incremental segmentation of a sensor space by the neural network which had a similar structure and function with Kohonen's self-organizing map. The incremental segmentation is done by adding neurons when a robot pays penalties. This technique provides incremental division of 'risky' areas within the sensor space. Chapman and Kaelbling [8] proposed G-learning that attempted to solve the "input generalisation problem" for the Q-learning algorithm. Munos and Patinel [9] proposed Partitioning Q-learning which rested basically on same principles as the G-learning. Both approaches are closely related to our work but aim at discrete sensor spaces only.

In this paper, we propose a Q-learning with adaptive state space construction. This is an advanced version of QLASS [10]. In the proposed algorithm, a robot starts with single state covering a whole sensor space. A new state is generated incrementally by segmenting a sub-region of the sensor space or combining existing states based on sensor vectors and reinforcement signals. A criterion for the incremental segmentation and combination is derived from the Q-learning algorithm. This is an online method to construct state space without *a priori* knowledge of an environment.

The next section gives a brief review of the reinforcement learning. The basic idea is then introduced, followed by an illustration of the learning algorithm. Some results of computer simulations are reviewed, and finally we conclude.

2   Reinforcement Learning

An N_s sensors robot provides N_s-dimensional vector $\mathbf{s}$, for which the i-th component is a range value s_i provided by the i-th sensor. For every sensor vector $\mathbf{s} \in \mathbf{S}$, the robot can take an action a from the action set $\mathbf{A}$. The action $a \in \mathbf{A}$ for the sensor vector $\mathbf{s} \in \mathbf{S}$ causes a transition of the sensor vector to $\mathbf{s}' = e(\mathbf{s}, a) \in \mathbf{S}$, where e is a given transition function which defines an environment. We assume that a fitness value $v(\mathbf{s})$ is defined for each sensor vector $\mathbf{s}$. The robot however can receive reinforcement signals only, which represent a fitness gain or loss between the sensor vector $\mathbf{s}$ and the next sensor vector $\mathbf{s}'$. Since the sensor vector $\mathbf{s}'$ is defined by a previous sensor vector $\mathbf{s}$ and an action a, we can define the reinforcement signal as $r(\mathbf{s}, a)$. A purpose of the reinforcement learning is to find a policy of selecting an action a for a sensor vector $\mathbf{s}$ that maximizes the discounted sum of the reinforcement signals $r(\mathbf{s}, a)$ received over time.

The Q-learning algorithm [11] gives us a sophisticated solution to this problem. A sensor space $\mathbf{S}$ is quantized into a discrete and finite set $\mathbf{X}$ as the Q-learning algorithm requires. A sub-region $\mathbf{S}^x$ of the sensor space $\mathbf{S}$ is characterised by a corresponding state $x \in \mathbf{X}$. If a sensor vector $\mathbf{s}$ is within this region, we say the robot is in the state x. An estimated discounted sum of the reinforcement signals $Q_n(x, a)$ for an action a is assigned to each state x, where n is the number of updates. If the robot is in a state x, an action a is selected from $\mathbf{A}$ according to Boltzmann distribution of a $Q_n(x, a)$ value, as follows,

$$P(a|x) = \frac{\exp(Q_n(x, a)/\tau)}{\sum_{b \in \mathbf{A}} \exp(Q_n(x, b)/\tau)} \tag{1}$$

where τ is a scaling constant. In the Q-learning algorithm, the robot transits from the state x to a state $x' \in \mathbf{X}$ by the action a and this updates the $Q_n(x, a)$ value as,

$$Q_{n+1}(x, a) \Leftarrow (1 - \alpha)Q_n(x, a) \\ + \alpha(r(\mathbf{s}, a) + \gamma \max_{b \in \mathbf{A}} Q_n(x', b)) \tag{2}$$

where α is a learning rate and γ is a discounting factor.

After a sufficient number of iteration, an action a which maximizes a $Q_n(x, a)$ is the optimal decision policy at a state x.

3   Basic Concept

The basic concept of the proposed algorithm is to find a state x in which a reinforcement signal $r(\mathbf{s}, a)$ for an action a is constant for every $\mathbf{s} \in \mathbf{S}^x$ and therefore we can define it as a function $r^x(a)$ of the action a. This is a method to model the fitness landscape by using inclined planes, as represented by Fig. 1.

If there is the only state x covering the whole sensor space $\mathbf{S}$ in which $r(\mathbf{s}, a)$ for an action a is constant for every $\mathbf{s} \in \mathbf{S}^x$, $Q_n(x, a)$ for the correct action a (here "correct action" means an action yielding the maximum reinforcement signal) can be calculated as follows,

$$Q_{n+1}(x, a) \Leftarrow (1 - \alpha)Q_n(x, a) \\ + \alpha(r_{max} + \gamma Q_n(x, a)) \tag{3}$$

where r_{max} is the maximum reinforcement signal given by,

$$r_{max} = \max_{\mathbf{s} \in \mathbf{S}, b \in \mathbf{A}} r(\mathbf{s}, b) \tag{4}$$

Equation 3 converges to,

$$Q_{max} = \frac{r_{max}}{1 - \gamma} \tag{5}$$

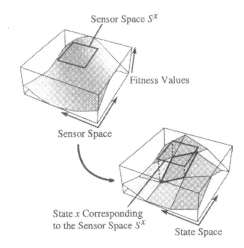

Sensor Space S^x

Fitness Values

Sensor Space

State x Corresponding
to the Sensor Space S^x

State Space

Fig. 1. QLASS defines a sub-space $\mathbf{S}^x$ as a collection of the sensor vectors $\mathbf{s}$ for which reinforcement signals $r(\mathbf{s}, a)$ is constant for any action a. The sub-space $\mathbf{S}^x$ is like a inclined plane with the gradient $r(\mathbf{s}, a)$ in a direction a.

In general, we need multiple states to cover the whole sensor space. We note here that the following condition is satisfied for every state x and number n,

$$\max_{b \in \mathbf{A}} Q_n(x, b) \leq Q_{max} \tag{6}$$

Thus, Eq. 2 can be transformed using $r^x(a)$,

$$
\begin{aligned}
Q_n(x, a) & \\
\leq & (1 - \alpha)Q_{n-1}(x, a) + \alpha(r^x(a) + \gamma Q_{max}) \\
= & (1 - \alpha)^n Q_0(x, a) \\
& + \alpha \sum_{i=0}^{n-1} (1 - \alpha)^i (r^x(a) + \gamma Q_{max}) \\
= & r^x(a) + \gamma Q_{max} \\
& + (1 - \alpha)^n (Q_0(x, a) - (r^x(a) + \gamma Q_{max}))
\end{aligned}
\tag{7}
$$

If the reinforcement signal $r^x(a)$ and the initial Q value $Q_0(x, a)$ satisfy the following conditions,

$$r^x(a) \geq 0, \quad Q_0(x, a) = 0 \tag{8}$$

we can obtain,

$$Q_n(x, a) \leq r^x(a) + \gamma Q_{max} \tag{9}$$

Consequently, we define a robot is in a same state x as long as the reinforcement signal $r(\mathbf{s}, a)$ satisfies the following equation under the conditions of Eq. 8,

$$Q_n(x, a) \leq r(\mathbf{s}, a) + \gamma Q_{max} \tag{10}$$

When $Q_n(x, a) - r(\mathbf{s}, a)$ exceeds γQ_{max}, a new state x' is added, for which the sensor vector $\mathbf{s}$ is within the sub-region $\mathbf{S}^{x'}$.

4 Learning Algorithm

We define each state x by a set $\mathcal{S}^x$ of representative sensor vectors $\mathbf{s}^x \in \mathcal{S}^x$. If $\mathbf{s}^x$ is the nearest vector of an incoming sensor vector $\mathbf{s}$, a robot is said to be in a state x. This criterion will result in a Voronoi tesselation of a sensor space. A procedure of the proposed algorithm using this definition of states is as follows:

1. [**Initialization**] We start with a single state x covering whole sensor space $\mathbf{S}$, which is defined by a set $\mathcal{S}^x = \{\mathbf{s}^x\}$ where $\mathbf{s}^x$ is a randomly intialized representative sensor vector. $Q_0(x, a) = 0$ is assigned to each action a.
2. [**Action based on Q values**] A robot with a sensor vector $\mathbf{s}$ in the state x is driven according to the Boltzmann distribution described in Eq. 1 and obtains the reinforcement signal $r(\mathbf{s}, a)$ as a result of an action a.
3. When the condition in Eq. 10 is not satisfied under condition of $n > \theta$, (a) otherwise (b). θ is a minimum updates of the state x for segmentation and combination.
   (a) [**Adding a new state**] A new state x' defined by the set $\mathcal{S}^{x'} = \{\mathbf{s}^{x'}\}$ is added for which the representative sensor vector $\mathbf{s}^{x'}$ is initialized to the sensor vector $\mathbf{s}$ of the robot. $Q_0(x', a)$ values for each action a is initialized to $Q_0(x', a) = 0$.
   (b) [**Updating Q value**] The $Q_n(x, a)$ value is updated according to Eq. 2 and the nearest representative sensor vector $\mathbf{s}^x$ to the sensor vector $\mathbf{s}$ is modified as follows
   $$\mathbf{s}^x \Leftarrow (1 - \beta)\mathbf{s}^x + \beta\mathbf{s} \tag{11}$$
   where β is the modification rate of the representative vector with $0 \leq \beta \leq 1$.
4. [**Combining states**] If $n > \theta$ is satisfied for all the $Q_n(x, a)$ value asigned to actions $a \in \mathbf{A}$, we calculate a distance $l_{x,y}$ between the state x and every other state y by
   $$l_{x,y} = \sqrt{\sum_{a \in \mathbf{A}} (Q(x, a) - Q(y, a))^2} \tag{12}$$
   When the distance $l_{x,y}$ satisfies the following condition for a small constant ε
   $$l_{x,y} < \varepsilon, \tag{13}$$
   we combine the set $\mathcal{S}^y$ to $\mathcal{S}^x$ as
   $$\mathcal{S}^x \Leftarrow \mathcal{S}^x \cup \mathcal{S}^y \tag{14}$$
   and remove $\mathcal{S}^y$.

Table 1. Parameters for simulations of navigating 2-sensors mobile robot.

parameters	description
$\tau = 1.0$	the scale constant (Eq. 1).
$\alpha = 0.5$	the learning rate (Eq. 2).
$\gamma = 0.3$	the discounting factor (Eq. 2).
$\beta = 0.1$	the modification rate of the representative vector (Eq. 11).
$\theta = 10$	the minimum updates of the states for segmentation and combination.
$\varepsilon = 0.1$	the small constant in Eq. 13.

5 Computer Simulations

5.1 Navigating 2-Sensors Mobile Robot

First, we applied the proposed algorithm to a problem navigating a mobile robot onto the top of a fitness landscape. We assume the mobile robot with $N_s = 2$ sensors which provide orthogonal coordinates of the robot as $\mathbf{s} = (d_0, d_1)$. The reinforcement signal $r(\mathbf{s}, a)$ is defined using the fintess value $v(\mathbf{s})$ as follows

$$r(\mathbf{s}, a) = \begin{cases} 0 & \text{if } v(\mathbf{s}') < v(\mathbf{s}) \\ 1 & \text{if } v(\mathbf{s}') = v(\mathbf{s}) \\ 2 & \text{if } v(\mathbf{s}') > v(\mathbf{s}) \end{cases} \tag{15}$$

The action set $\mathbf{A}$ is defined as follows

$$\mathbf{A} = \{stop, up, down, right, left\} \tag{16}$$

for which the environmental function $e(\mathbf{s}, a)$ is

$$e(\mathbf{s}, a) = \mathbf{s} + d \cdot \mathbf{u}_a \tag{17}$$

with a constant $d = 0.005$ and vectors $\mathbf{u}_a$ are defined as follows

$$\mathbf{u}_a = \begin{cases} (0,0) & \text{for } a = stop \\ (0,1) & \text{for } a = up \\ (0,-1) & \text{for } a = down \\ (1,0) & \text{for } a = right \\ (-1,0) & \text{for } a = left \end{cases} \tag{18}$$

Other parameters for simulations are summarized in Table 1.

Two configurations of the fitness landscape shown in Fig. 2 are used for computer simulations. The lighter areas correspond to the higher fitness values and the darker areas to the lower. They are generated by the following equations for every sensor vector $\mathbf{s} = (d_0, d_1)$,

$$v(\mathbf{s}) = \begin{cases} \cos(d_0) \cdot \sin(d_1) & \text{configuration (i)} \\ \sin(d_0 \cdot d_1) & \text{configuration (ii)} \end{cases} \tag{19}$$

Table 2. Q values for the major states for each configurations.

For the configuration (i).

state	actions				
number	stop	up	down	right	left
58	1.81	2.08	1.17	1.00	2.09
425	1.84	2.79	0.41	2.85	0.71
463	1.86	1.35	2.63	2.82	1.32
1408	1.86	0.79	2.68	0.62	2.86

For the configuration (ii).

state	actions				
number	stop	up	down	right	left
80	1.86	2.86	0.86	2.86	0.86
169	1.39	0.36	2.86	0.59	2.44

A procedure of the simulation for the each configuration is as follows:

1. Place a robot with single state at randomly chosen coordinates.
2. Apply the proposed algorithm to the robot for 500 steps till it reaches the top of the fitness landscape where $v(s) = 1.0$. We count this 1 trial.
3. Replace the robot at randomly chosen coordinates and repeat trials.

Figure 3 shows time courses of the number of states and success rate. The success rate indicates how many times the robot can reach the top of the fitness landscape in last 10 trials. Figure 4 shows resulting state spaces after 100 trials. Solid lines in the diagrams represent boundaries of sub-regions covered by representative sensor vectors. Each dot in the sub-region corresponds to the representateive sensor vector s^x. Note that each sub-region does not correspond to a state. Since each state is defined by a set of representative vectors, a state can cover multiple sub-regions.

Figure 5 shows the number of the representative sensor vectors included in each state. For the configuratoin (i), almost all representative sensor vectors are included in only 4 states. Similarly, almost all representative sensor vectors of the configuration (ii) are included in only 2 states. Figure 6 shows the representateive sensor vectors included in the 4 major states numbered for the configuration (i). Figure 7 shows the representative sensor vectors included in the 2 major states of the configuration (ii). Their Q values are summarized in Table 2.

It is clear that the robot can reach the top of the fitness landscape by using the obtained state space and the Q values. Furthermore, fine segmentation of the sensor space is done by fine assignment of the representative sensor vectors around the top of the fitness landscape where are important to accomplish the task and they are combined into small number of states. The obtained state space is free from the "perceptual aliasing problem".

5.2 Navigating 6-Sensors Mobile Robot

We applied the proposed algorithm to realistic problem navigating a mobile robot with a similar structure to real robot "Khepera" [12]. As shown in Fig. 8 (a), we assume that the mobile robot is 100mm in diameter and has $N_s = 6$ sensors measuring distances between the edge of the robot and obstacles. They provide a 6-dimensional sensor vector s. The robot can move constant distance $l = 50mm$ in four direction; straight forward, right forward, left forward and straight backward. Then, the action set **A** is defined by

$$\mathbf{A} = \{stop, forward, right, left, backward\} \tag{20}$$

A task is navigating the robot from initial position to the goal in a simple maze shown in Fig. ?? (b). In this case, we cannot define the environmental function and the fitness landscape explicitly. We define only the reinforcement signals for every situations as follows

$$r(\mathbf{s}, a) = \begin{cases} 0 & \text{if the robot collides with the wall} \\ 2 & \text{if the robot reaches the goal area} \\ 1 & \text{otherwise.} \end{cases} \tag{21}$$

Other parameters for simulations are summarized in Table 3. Since the maximum reinforcement signal is given only when the robot reaches the goal area, the discounting rate γ is a little larger than that in Table 1.

A procedure of the simulation is as follows:

1. Place a robot with single state Q values at initial position in the simple maze.
2. Apply the proposed algorithm to the robot for 1000 steps till it reaches the goal. We count this 1 trial.
3. Replace the robot at randomly chosen neighborhood of the initial position and repeat trials.

Figure 9 shows time courses of the number of states and success rate. As the number of states increases, the robot learns a way to the goal.

Figure 10 shows trajectories of the robot in different trials. Both robots in the trial 250 and the trial 500 can reach the goal, but the trajectory of the latter is rather smooth than that of the former. The robot of the trial 250 takes 910 steps and that of trial 500 takes 190 steps to reach the goal.

Figure 11 shows the projection of the representative sensor vectors at trial 500 onto forward, left and right sensor space. A dense area around forward sensor $= 200$ and left sensor $= 0$ to 500 (surrounded by solid lines in Fig. 11) corresponds to right-lower corner and right-upper corner in Fig. 10. Obviously, actions around here in the maze is important to accomplish the task. This is reflected in the state space Fig. 11.

6 Conclusions and Future Directions

This paper presented a combination of Q-learning and adaptive construction of a state space based on reinforcement signals. Compared with other approaches, the proposed algorithm is realized in a simple way and is nearly free from determining parameters which are used for judgement in segmenting and combining state

Table 3. Parameters for simulations of navigating 6-sensors mobile robot in the simple maze

parameters	description
$\tau = 1.0$	the scale constant (Eq. 1)
$\alpha = 0.5$	the learning rate (Eq. 2)
$\gamma = 0.5$	the discounting factor (Eq. 2)
$\beta = 0.1$	the modification rate of the representative vector (Eq. 11)
$\theta = 5$	the minimum updates of the states for segmentation and combination.
$\varepsilon = 0.1$	the small constant in Eq. 13.

space. In the proposed algorithm, the judgement parameter γQ_{max} is calculated from the maximum reinforcement signal r_{max} and the discounting factor γ.

The proposed algorithm was applied to the problems navigating the mobile robot in continuous sensor space. Simulation results showed that it could generate a discrete state space efficiently to accomplish these tasks without *a priori* knowledge of environments. Resulting distribution of representative sensor vectors revealed the fitness landscape in a Voronoi tessellation.

Finally, we discuss the remaining problems and future works:

- *Stochastic environments.* In a real environment, same actions for a sensor vector generate different sensor vectors. Such environment can be described as $e(\mathbf{s}, a, \chi)$ using a probabilistic variable χ. This causes difficulties in determining boundaries of states. To cope with this problem, we should introduce a probabilistic criterion for adding a new state to the proposed algorithm.
- *Dynamic environments.* In an environment including active elements such as other robots, it is important to estimate dynamics of the active elements from fluctuations of sensor vectors. In this case, we can introduce a mehtod of function approximation such as an artificial neural network preliminary to the proposed algorithm.
- *Incomplete perception.* Even in a deterministic environments, the incomplete perception[13] causes the "perceptual aliasing problem". A simple solution is to introduce a short-term memory and define a state by using a set of current and past sensor vectors, for which we should extend the proposed algorithm.

Acknowledgment

This research was supported by the Grand-in-Aid for Scientific Reseach on Priority Area No.264 from the Ministry of Education, Science, Sports, and Culture of Japan.

References

1. Whitehead, S. D., Ballard, D. H.: Learning to perceive and act by trial and error. ML **7** (1991) 45–83
2. Chrisman, L.: Reinforcement Learning with Perceptual Aliasing: The Perceptual Distinctions Approach. Proc. of AAAI-92 (1992) 183–188
3. Tan, M.: Cost-sensitive reinforcement learning for adaptive classification and control. Proc. of AAAI-91 (1991)
4. Kröse, B.J.A., van Dam, J.W.M.: Adaptive state space quantisation for reinforcement learning of collision-free navigation. Proc. of IROS **2** (1992) 1327–1331
5. Dubrawski, A., Reignier, P.: Learning to Categorize Perceptual Space of a Mobile Robot Using Fuzzy-ART Neural Network. Proc. of IROS **2** (1994) 1272–1277
6. Takahashi, Y., Asada, M., Hosoda, K.: Reasonable Performance in Less Learning Time by Real Robot Based on Incremental State Space Segmentation. Proc. of IROS **3** (1996) 1518–1524
7. Ishiguro, H., Sato, R., Ishida, T.: Robot Oriented State Space Construction. Proc. of IROS **3** (1996) 1496–1501.
8. Chapman, D., Kaebling, L. P.: Input Generalisation in Delayed Reinforcement Learning: an Algorithm and Performance Comparisons. Proc. of IJCAI-91 (1991) 726–731
9. Munos, R., Patinel, J.: Reinforcement learning with dynamic covering of state-action space: Partitioning Q-learning. Proc. of SAB (1994) 354–363
10. Murao, H., Kitamura, S.: Q-Learning with Adaptive State Segmentation (QLASS). Proc. of CIRA (1997) 179–184
11. Watkins, C.: Learning from Delayed Rewards. Ph.D. Dissertation of Cambridge University (1989)
12. Naito, T., Odagiri, R., Matsunaga, Y., Tanifuji, M., Murase, K.: Genetic evolution of a logic circuit which controls an autonomous mobile robot. Proc. of ICES96 (1996)
13. MaCallum, R. A.: Overcoming Incomplete Perception with Utile Distinction Memory. Proc. of ICML (1993) 190–196

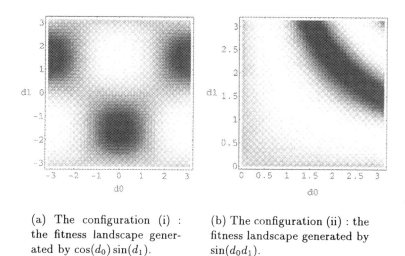

(a) The configuration (i) : the fitness landscape generated by $\cos(d_0)\sin(d_1)$.

(b) The configuration (ii) : the fitness landscape generated by $\sin(d_0 d_1)$.

Fig. 2. Two configurations of the fitness landscape. The lighter areas correspond to the higher fitness values and the darker areas to the lower.

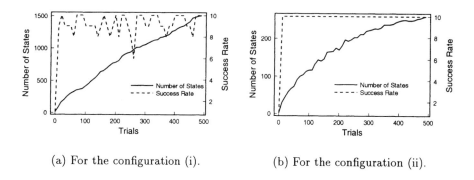

(a) For the configuration (i).

(b) For the configuration (ii).

Fig. 3. Time courses of the number of the states and success rate. The success rate indicates how many times the robot can reach the top of the fitness landscape in 10 trials.

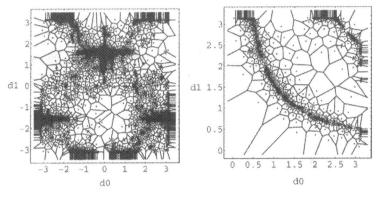

(a) For the configuration (i).   (b) For the configuration (ii).

Fig. 4. Resulting state space after 100 trials. Solid lines represent boundaries of sub-regions covered by representative sensor vectors. Each dot in the sub-region corresponds to the representative sensor vector $s(x)$. Fine segmentation is done in important areas to accomplish the task.

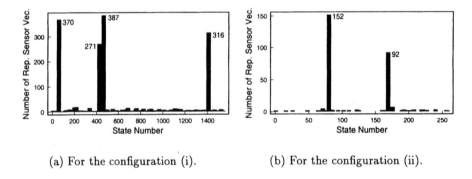

(a) For the configuration (i).   (b) For the configuration (ii).

Fig. 5. The number of the representative sensor vectors included in the each state. For the configuration (i), states 58, 425, 463 and 1408 consist of 370, 271, 387 and 316 representative sensor vectors, respectively. For the configuration (ii), states 80 and 169 consist of 152 and 92 representative sensor vectors, respectively.

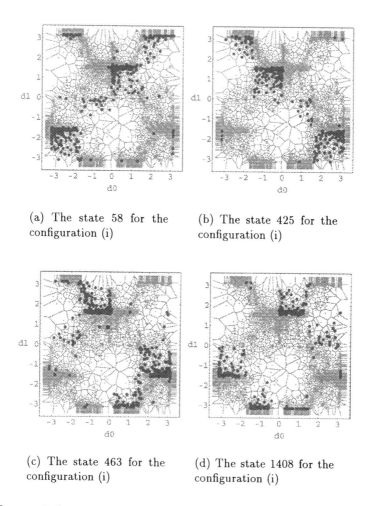

(a) The state 58 for the configuration (i)

(b) The state 425 for the configuration (i)

(c) The state 463 for the configuration (i)

(d) The state 1408 for the configuration (i)

Fig. 6. Representative sensor vectors included in the 4 major states 58, 425, 463 and 1408 for the configuration (i) are represented in large dots. These states complement each other and cover almost the whole sensor space.

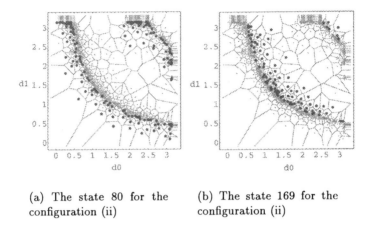

(a) The state 80 for the configuration (ii)

(b) The state 169 for the configuration (ii)

Fig. 7. Representative sensor vectors included in the 2 major states 80 and 169 are represented in large dots. These states complement each other and cover almost the whole sensor space.

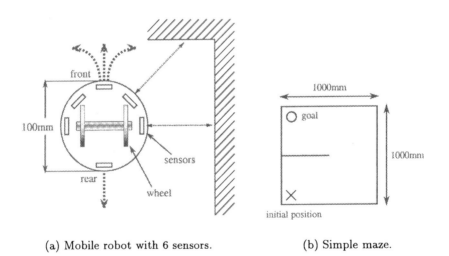

(a) Mobile robot with 6 sensors.

(b) Simple maze.

Fig. 8. A mobile robot for simulation studies with 6 sensors which measure distances between the edge of the robot and obstacles. The robot can move constant distance at single instance in 4 direction, straight forward, right forward, left forward and straight backward. The task is navigating the mobile robot from initial position to the goal in a simple maze shown in (b).

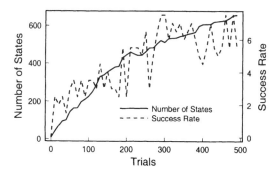

Fig. 9. Time courses of the number of states and success rate. The success rate indicates how many times the robot can reach the top of the fitness landscape in 10 trials.

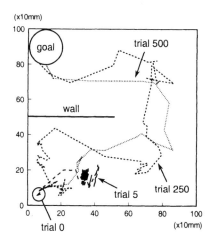

Fig. 10. Trajectories of the robot in different trials. The robot can reach the goal in 910 steps at trial 250 and in 190 steps at trial 500.

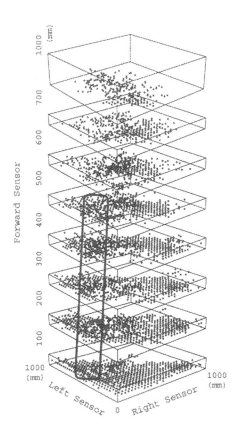

Fig. 11. A projection of the representative sensor vectors at trial 500 onto forward, left and right sensor space. A surrounded area corresponds to the right-lower corner and right-upper corner in Fig. 10. The denseness in the area reflects the importance of corresponding sensor vectors to accomplish the task.

Modular Reinforcement Learning: An Application to a Real Robot Task

Zsolt Kalmár[1], Csaba Szepesvári[2], and András Lőrincz[3]

[1] Dept. of Informatics JATE, Szeged, Aradi vrt. tere 1, Hungary, H-6720
{kalmar,szepes,lorincz}@iserv.iki.kfki.hu
[2] Research Group on Art. Int., JATE, Szeged, Aradi vrt. tere 1, Hungary, H-6720
[3] Dept. of Chemical Physics, Inst. of Isotopes, HAS, Budapest, P.O. Box 77, Hungary, H-1525

Abstract. The behaviour of reinforcement learning (RL) algorithms is best understood in completely observable, finite state- and action-space, discrete-time controlled Markov-chains. Robot-learning domains, on the other hand, are inherently infinite both in time and space, and moreover they are only partially observable. In this article we suggest a systematic design method whose motivation comes from the desire to transform the task-to-be-solved into a finite-state, discrete-time, "approximately" Markovian task, which is completely observable too. The key idea is to break up the problem into subtasks and design controllers for each of the subtasks. Then operating conditions are attached to the controllers (together the controllers and their operating conditions which are called modules) and possible additional features are designed to facilitate observability. A new discrete time-counter is introduced at the "module-level" that clicks only when a change in the value of one of the features is observed. The approach was tried out on a real-life robot. Several RL algorithms were compared and it was found that a model-based approach worked best. The learnt switching strategy performed equally well as a handcrafted version. Moreover, the learnt strategy seemed to exploit certain properties of the environment which could not have been seen in advance, which predicted the promising possibility that a learnt controller might overperform a handcrafted switching strategy in the future.

1 Introduction

Reinforcement learning (RL) is the process of learning the coordination of concurrent behaviours and their timing. A few years ago Markovian Decision Problems (MDPs) were proposed as the model for the analysis of RL [17] and since then a mathematically well-founded theory has been constructed for a large class of RL algorithms. These algorithms are based on modifications of the two basic dynamic-programming algorithms used to solve MDPs, namely the value- and policy-iteration algorithms [25, 5, 10, 23, 18]. The RL algorithms learn via experience, gradually building an estimate of the optimal value-function, which is known to encompass all the knowledge needed to behave in an optimal way according to a fixed criterion, usually the expected total discounted-cost criterion.

The basic limitations of all of the early theoretical results of these algorithms was that they assumed finite state- and action-spaces, and discrete-time models in which the state information too was assumed to be available for measurement. In a real-life problem however, the state- and action-spaces are infinite, usually non-discrete, time is continuous and the system's state is not measurable (i.e. with the latter property the process is only partially observable as opposed to being completely observable). Recognizing the serious drawbacks of the simple theoretical case, researchers have begun looking at the more interesting yet theoretically more difficult cases (see e.g. [11, 16]). To date, however, no complete and theoretically sound solution has been found to deal with such involved problems. In fact the above-mentioned learning problem is indeed intractable owing to partial-observability. This result follows from a theorem of Littman's [9].

In this article an attempt is made to show that RL can be applied to learn real-life tasks when *a priori* knowledge is combined in some suitable way. The key to our proposed method lies in the use of high-level modules along with a specification of the operating conditions for the modules and other "features", to transform the task into a finite-state and action, completely-observable task. Of course, the design of the modules and features requires a fair amount of *a priori* knowledge, but this knowledge is usually readily available. In addition to this, there may be several possible ways of breaking up the task into smaller subtasks but it may be far from trivial to identify the best decomposition scheme. If all the possible decompositions are simultaneously available then RL can be used to find the best combination. In this paper we propose design principles and theoretical tools for the analysis of learning and demonstrate the success of this approach via *real-life* examples. A detailed comparison of several RL methods, such as Adaptive Dynamic Programming (ADP), Adaptive Real-Time Dynamic Programming (ARTDP) and Q-learning is provided, having been combined with different exploration strategies.

The article is organized in the following way. In the next section (Section 2) we introduce our proposed method and discuss the motivations behind it. In it the notion of "approximately" stationary MDPs is also introduced as a useful tool for the analysis of "module-level" learning. Then in Section 3 the outcome of certain experiments using a mobile robot are presented. The relationship of our work to that of others is contrasted in Section 4, then finally our conclusions and possible directions for further research are given in Section 5. Due to the lack of space some details were left out from this article, but these can be found in [8].

2 Module-based Reinforcement Learning

First of all we will briefly run through Markovian Decision Problems (MDPs), a value-function approximation-based RL algorithm to learn solutions for MDPs and their associated theory. Next the concept of recursive-features and time discretization based on these features are elaborated upon. This is then followed by a sensible definition and principles of module-design together with a brief explanation of why the modular approach can prove successful in practice.

2.1 Markovian Decision Problems

RL is the process by which an agent improves its behaviour from observing its own interactions with the environment. One particularly well-studied RL scenario is that of a single agent minimizing the expected-discounted total cost in a discrete-time finite-state, finite-action environment, when the theory of MDPs can be used as the underlying mathematical model. A finite MDP is defined by the 4-tuple (S, A, p, c), where S is a finite set of states, A is a finite set of actions, p is a matrix of transition probabilities, and c is the so-called immediate cost-function. The ultimate target of learning is to identify an optimal policy. A policy is some function that tells the agent which set of actions should be chosen under which circumstances. A policy π is optimal under the *expected discounted total cost criterion* if, with respect to the space of all possible policies, π results in a minimum expected discounted total cost for all states. The optimal policy can be found by identifying the optimal value-function, defined recursively by

$$v^*(s) = \min_{a \in U(s)} \left(c(s, a) + \gamma \sum_{s'} p(s, a, s') v^*(s') \right)$$

for all states $s \in S$, where $c(s, a)$ is the immediate cost for taking action a from state s, γ is the discount factor, and $p(s, a, s')$ is the probability that state s' is reached from state s when action a is chosen. $U(s)$ is the set of admissible actions in state s. The policy which for each state selects the action that minimizes the right-hand-side of the above fixed-point equation constitutes an optimal policy. This yields the result that to identify an optimal policy it is sufficient just to find the optimal value-function v^*. The above simultaneous non-linear equations (non-linear because of the presence of the minimization operator), also known as the *Bellman equations* [3], can be solved by various dynamic programming methods such as the value- or policy-iteration methods [15].

RL algorithms are generalizations of the DP methods to the case when the transition probabilities and immediate costs are unknown. The class of RL algorithms of interest here can be viewed as variants of the value-iteration method: these algorithms gradually improve an estimate of the optimal value-function via learning from interactions with the environment. There are two possible ways to learn the optimal value-function. One is to estimate the model (i.e., the transition probabilities and immediate costs) while the other is to estimate the optimal action-values directly. The optimal action-value of an action a given a state s is defined as the total expected discounted cost of executing the action from the given state and proceeding in an optimal fashion afterwards:

$$Q^*(s, a) = c(s, a) + \gamma \sum_{s'} p(s, a, s') v^*(s'). \tag{1}$$

The general structure of value-function-approximation based RL algorithms is given in Table 1. In the RL algorithms various models are utilized along with

Initialization: Let $t = 0$, and initialize the utilized model (M_0) and the Q-function (Q_0)

Repeat Forever

1. Observe the next state s_{t+1} and reinforcement signal c_t .

2. Incorporate the new experience (s_t, a_t, s_{t+1}, c_t) into the model and into the estimate of the optimal Q-function: $(M_{t+1}, Q_{t+1}) = F_t(M_t, Q_t, (s_t, a_t, s_{t+1}, c_t))$.

3. Choose the next action to be executed based on (M_{t+1}, Q_{t+1}): $a_{t+1} = S_t(M_{t+1}, Q_{t+1}, s_{t+1})$ and execute the selected action.

4. $t := t + 1$.

Table 1. The structure of value-function-approximation based RL algorithms.

an update rule F_t and action-selection rule S_t.

In the case of the Adaptive Real-Time Dynamic Programming (ARTDP) algorithm the model consists (M_t) of the estimates of the transition probabilities and costs, the update-rule F_t being implemented e.g. as an averaging process. Instead of the optimal Q-function, the optimal value-function is estimated and stored to spare storage space, and the Q-values are then computed by replacing the true transition probabilities, costs and the optimal value-function in Equation 1 by their estimates. An update of the estimate for the optimal value-function is implemented by an asynchronous dynamic programming algorithm using an inner-loop in Step 2 of the algorithm.

Another popular RL algorithm is Q-learning, which does not employ a model but instead the Q-values are updated directly according to the iteration procedure [25]

$$Q_{t+1}(s_t, a_t) = (1 - \alpha_t(s_t, a_t))Q_t(s_t, a_t) +$$

$$\alpha_t(s_t, a_t)(c_t + \gamma \min_a Q_t(s_{t+1}, a)),$$

where $\alpha_t(s_t, a_t) \geq 0$, and satisfies the usual Robins-Monro type of conditions. For example, one might set $\alpha_t(s, a) = \frac{1}{n_t(s,a)}$ but often in practice $\alpha_t(s, a) =$ const is employed which while yielding increased adaptivities no longer ensures convergence.

Both algorithms mentioned previously are guaranteed to converge to the optimal value-/Q-function if each state-action pair is updated infinitely often. The action selection procedure S_t should be carefully chosen so that it fits the dynamics of the controlled process in a way that the condition is met. For example, the execution of random actions meets this "sufficient-exploration" condition when the MDP is ergodic. However, if on-line performance is important then

more sophisticated exploration is needed which, in addition to ensuring sufficient exploratory behaviour, exploits accumulated knowledge. For more details
the interested reader is referred to [8].

2.2 Recursive Features and Feature-based Time-Discretization

In case of a real-life robot-learning task the dynamics cannot be formulated
exactly as a *finite* MDP, nor is the state information available for measurement.
This latter restriction is modelled by Partially-Observable MDPs (POMDPs)
where (in the simplest case) one extends an MDP with an *observation function*
h which maps the set of states S into a set X, called the observation set (which
is usually non-countable, just like S). The defining assumption of a POMDP is
that the full state s can be observed only through the observation function, i.e.
only $h(s)$ is available as input and this information alone is usually insufficient
for efficient control since h is usually a non-injection (i.e. h may map different
states to the same observations). *Features* which mathematically are just well-
designed observation functions, are known to be efficient in dealing with the
problem of infinite state-spaces. Moreover, when their definitions are extended
in a sensible way they become efficient in dealing with partial observability.

It is well known that POMDPs optimal policies can depend on their whole
past histories. This leads us onto a mere generalization of features, such that
the feature's values can depend on all past observations, i.e. mathematically a
feature becomes an infinite sequence of mappings $(f^0, f^1, \ldots f^t, \ldots)$, with $f^t :
(X \times A)^t \times X \to F$, where F and X are the feature- and observation-spaces. Since
RL is supposed to work on the output of features, and RL requires finite spaces it
means that F should be finite. Features that require infinite memory are clearly
impossible to implement, so features used in practice should be restricted in
such a way that they require a finite "memory". For example, besides stationary
features which take the form $(f^0, f^0, \ldots, f^0, \ldots)$ (i.e. $f^t = f^0$ for all $t \geq 1$) and
are called *sensor-based* features, *recursive* features (in control theory these are
called *filters* or *state-estimators*) are those that can be implemented using a finite
memory[1]. For example, in the case of a one-depth recursive feature the value at
the tth step is given by $f_t = R(x_t, a_{t-1}, f_{t-1})$, where $R : X \times A \times F \to F$ defines
the recursion and $f_0 = f^0(x_0)$ for some function $f^0 : X \to F$.[2] Features whose
values depend on the past-observations of a finite window form a special class of
recursive filters. Later in the application-section some example features will be
presented.

Instead of relying on a single feature, it is usually more convenient to define
and employ a set of features, each of which indicates a certain event of interest.
Note that a finite set of features can always be replaced by a single feature whose

---

[1] If the state space is infinite then not all sensor-based features can be realized in
practice.

[2] If time is continuous then recursive features should be replaced by features that admit
continuous-time dynamics. Details concerning continous time systems are given in
[8].

output-space is the Cartesian product of the output spaces of the individual features and whose values can be computed componentwise by the individual single features. That is to say, the new feature's values are the 'concatenated' values of the individual features.

Since the feature-space is finite, a natural discretization of time can be obtained. The new time-counter clicks only when the feature value jumps in the feature-space. This makes it useful to think of such features as event-indicators which represent the actuality of certain conditions of interest. This interpretation gives us an idea of how to define features in such a way that the dynamics at the level of the new counter are simplified.

2.3 Modules

So far we have realized that the new "state-space" (the feature-space) and "time" can be made discrete. However, the action-space may still be infinite. Uniform discretization which lacks a priori knowledge would again be impractical in most of the cases (especially when the action space is unbounded), so we would rather consider an idea motivated by a technique which is often applied to solve large search-problems efficiently. The method in question divides the problem into smaller subproblems which are in turn divided into even smaller sub-problems, etc., then at the end routines are provided that deal with the resulting mini-problems. The solution of the entire problem is then obtained by working backwards: At every moment the mini-routine corresponding to the actual state of the search problem is applied. To put it in another way, the problem-solver defines a set of sub-goals, sub-sub-subgoals, etc. in such a way that if one of the sub-goals is satisfied then the resolution of the main-goal will be easy to achieve. In control-tasks the same decomposition can usually be done with respect to the main control objective without any difficulty. The routines that resolve the very low-level subgoals should be provided by closed-loop controllers which achieve the given subgoal under the set conditions (the conditions are usually given implicitly, e.g. the condition is fulfilled if the "predecessor" subgoals have already been achieved). The process of breaking up the problem into small subtasks can be repeated several times before the actual controllers are designed, so that the complexity of the individual controllers can be kept low. The controllers together with their operating conditions, which may serve as a basic set of features, will be called modules. In principle a consistent transfer of the AI-decomposition yields the result that the operating conditions of the situation are exclusive and cover every situation. However, such a solution would be very sensitive to perturbations and unmodelled dynamics.

A more robust solution can be obtained by extending the range of operating conditions and may mean that more than one controller can be applied at the same time. This calls for the introduction of a mechanism (the switching function) which determines which controller has to be activated if there are more than one available. More specifically, in our case a switching function S maps feature-vectors (which are composed of the concatenation of the operating conditions of the individual modules and some possible additional features) to the indeces of modules and at any time the module with index $S(f)$ where f id observed feature vector is activated. Of course only those modules can be activated whose operating conditions are satisfied. The operation of the whole mechanism is then the following. A controller remains active until the switching function switches to another controller. Since the switching function depends on the observed feature-values the controllers will certainly remain active till a change in the feature-vector is observed. We further allow the controllers to manipulate the own observation-process. In this the controllers may inhibit an observation from occurring and thus may hold up their activity for a while.

The goal of the design procedure is to set up the modules and additional features in such a way that there exists a proper switching controller $S : F \to \{1, 2, \ldots, n\}$ which for any given history results in a closed-loop behaviour which fulfills the "goal" of control in time. It can be extremely hard to prove even the *existence* of such a valid switching controller. One approach is to use a so-called *accessibility decision problem* for this purpose which is a discrete graph with its node set being in our case the feature set F and the edges connect features which can be observed in succession. Then standard DP techniques can be used to decide the existence of a proper switching controller [8].

Of course, since the definitions of the modules and features depend on the designer, it is reasonable to assume that by clever design a satisfactory decomposition and controllers could be found even if only qualitative properties of the controlled object were known. RL could then be used for two purposes: either to find the best switching function assuming that at least two proper switching functions exist, or to decide empirically whether a valid switching controller exists at all. The first kind of application of RL arises as result of the desire to guarantee the existence of a proper switching function through the introduction of more modules and features than is minimally needed. But then good switching which exploits the capabilities of all the available modules could well become complicated to find manually.

If the accessibility decision problem were extendible with transition-probabilities to turn it to an MDP [3] then RL could be rightly applied to find the best switching function. For example if one uses a fixed (maybe stochastic) stationary switching policy and provided that the system dynamics can be formulated as an MDP then there is a theoretically well-founded way of introducing transition-probabilities (see [16]). Unfortunately, the resulting probabilities may well depend on the switching policy which can prevent the convergence of the RL algorithms. However, the following "stability" theorem shows that the difference of the cost of optimal policies corresponding to different transition probabilities is proportional to the extent the transition probabilities differ, so we may expect that a slight change in the transition probabilities does not result in completely different optimal switching policies and hence, as will be explained shortly after the theorem, we may expect RL to work properly, after all.

---

[3] Note that as the original control problem is deterministic it is not immediate when the introduction of probabilities can be justified. One idea is to refer to the ergodicity of the control problem.

Theorem 21 *Assume that two MDPs differ only in their transition-probability matrices, and let these two matrices be denoted by p_1 and p_2. Let the corresponding optimal cost-functions be v_1^* and v_2^*. Then*

$$\|v_1^* - v_2^*\| \leq \gamma \frac{nC\|p_1 - p_2\|}{(1-\gamma)^2},$$

where $C = \|c\|$ is the maximum of the immediate costs, $\|\cdot\|$ denotes the supremum-norm and n is the size of the state-space.

Proof. Let T_i be the optimal-cost operator corresponding to the transition-probability matrix p_i, i.e.

$$(T_i v)(s) = \min_{a \in U(x)} \left(c(s,a) + \gamma \sum_{s' \in X} p_i(s,a,s')v(s') \right),$$

$$v : S \to \Re, \, i = 1, 2.$$

Proceeding with standard fixpoint and contraction arguments (see e.g. [19]) we get that $\|v_1^* - v_2^*\| \leq \|T_1 v_1^* - T_1 v_2^*\| + \|T_1 v_2^* - T_2 v_2^*\|$ and since T_1 is a contraction with index γ, and the inequality $\|T_1 v - T_2 v\| \leq \gamma \|p_1 - p_2\| \sum_{y \in X} |v(y)|$ we obtain $\delta = \|v_1^* - v_2^*\| \leq \gamma\delta + \gamma \|p_1 - p_2\| |X| C/(1-\gamma)$, where $\|v_1^*\| \leq C/(1-\gamma)$ has been employed [15]. Rearranging the inequality in terms of δ then yields Theorem 21.

Motivated by the previous theorem we define ε-stationary MDPs as the quadruple (S, A, p, c), where S, A and c are as before but p, the transition probability matrix, may vary in time but with $\|p_t - p^*\| \leq \varepsilon$ holding for all $t > 0$. Our expectations are that although the transitions cannot be modelled with a fixed transition probability matrix (i.e. stationary MDP), they can be modelled by an ε-stationary one even if the switching functions are arbitrarily varied and we conjecture that RL methods would then result in oscillating estimates of the optimal value-function, but with the oscillation being asymptotically proportional to ε. Note that ε-stationarity was clearly observed in our experiments which we will describe now.

3 Experiments

The validity of the proposed method was checked with actual experiments carried out using a Khepera-robot. The robot, the experimental setup, general specifications of the modules and the results are all presented in this section.

3.1 The Robot and its Environment

The mobile robot employed in the experiments is shown in Figure 1. It is a Khepera[4] robot equipped with eight IR-sensors, six in the front and two at the

---

[4] The Khepera was designed and built at Laboratory of Microcomputing, Swiss Federal Institute of Technology, Lausanne, Switzerland.

Fig. 1. The Khepera and the experimental environment. The task was to grasp a ball and hit the stick with it.

back, the IR-sensors measuring the proximity of objects in the range 0-5 cm. The robot has two wheels driven by two independent DC-motors and a gripper which has two degrees of freedom and is equipped with a resistivity sensor and an object-presence sensor. The vision turret is mounted on the top of the robot as shown. It is an image-sensor giving a linear-image of the horizontal view of the environment with a resolution of 64 pixels and 256 levels of grey. The horizontal viewing-angle is limited to about 36 degrees. This sensor is designed to detect objects in front of the robot situated at a distance spanning 5 to 50 cm. The image sensor has no tilt-angle, so the robot observes only those things whose height exceeds 5 cm.

The learning task was defined as follows: find a ball in an arena, bring it to one of the corners marked by a stick and hit the stick with the ball. The robot's environment is shown in Figure 1. The size of the arena was 50 cm x 50 cm with a black coloured floor and white coloured walls. The stick was black and 7 cm long, while three white-coloured balls with diameter 3.5 cm were scattered about in the arena. The task can be argued to have been biologically inspired because it can be considered as the abstraction of certain foraging tasks or a "basketball game". The environment is highly chaotic because the balls move in an unpredictable manner and so the outcome of certain actions is not completely predictable, e.g. a grasped ball may easily slip out from the gripper.

3.2 The Modules

Subtask decomposition Firstly, according to the principles laid down in Section 2, the task was decomposed into subtasks. The following subtasks were naturally: (T1) to find a ball, (T2) grasp it, (T3) bring it to the stick, and (T4) hit the stick with the grasped ball. Subtask (T3) was further broken into two subtasks, that of (T3.1) 'safe wandering' and (T3.2) 'go to the stick', since the robot cannot see the stick from every position and direction. Similarly, because of the robot's limited sensing capabilities, subtask (T1) was replaced by safe-wandering and subtask (T2) was refined to 'when an object nearby is sensed examine it and grasp it if it is a ball'. Notice that subtask 'safe wandering' is used for two purposes (to find a ball or the stick). The operating conditions of the corresponding controllers arose naturally as (T2) – an object should be nearby, (T3.2) – the stick should be detected, (T4) – the stick should be in front of the robot, and (T1,T3.1) – no condition. Since the behaviour of the robot must differ before and after locating a ball, an additional feature indicating when a ball was held was supplied. As the robot's gripper is equipped with an 'object-presence' sensor the 'the ball is held" feature was easy to implement. If there had not been such a sensor then this feature still could have been implemented as a switching-feature: the value of the feature would be 'on' if the robot used the grasping behaviour and hence not the hitting behaviour. An 'unstuck' subtask and corresponding controller were also included since the robot sometimes got stuck. Of course yet another feature is included for the detection of "goal-states". The corresponding feature indicates when the stick was hit by the ball. This feature's value is 'on' iff the gripper is half-closed but the object presence sensor does not give a signal. Because of the implementation of the grasping module (the gripper was closed only after the grasping module was executed) this implementation of the "stick has been hit by the ball" feature was satisfactory for our purposes, although sometimes the ball slipped out from the gripper in which case the feature turned 'on' even though the robot did not actually reach the goal. Fortunately this situation did not happen too often and thus did not affect learning.

The resulting list of modules and features is shown in Table 2. The controllers work as intended, some fine details are discussed here (for more complete description see [8]). For example, the observation process was switched off until the controller of Module 3was working so as the complexity of the module-level decision problem is reduced. The dynamics of the controller associated with Module 1 were based on the maximization of a function which depended on the proximity of objects and the speed of both motors[5]. If there were no obstacles near the robot this module made the robot go forward. This controller could thus serve as one for exploring the environment. Module 2 was applicable only if the stick was in the viewing-angle of the robot, which could be detected in an unambiguous way because the only black thing that could get into the view of the robot was the stick. The range of allowed behaviour associated with this module was implemented as a proportional controller which drove the robot in such a way that the angle difference between the direction of motion and line of sight to the stick was reduced. The behaviour associated with Module 3 was applicable only if there was an object next to the robot, which was defined as a function of the immediate values of IR-sensors. The associated behaviour was the following: the robot turned to a direction which brought it to point directly at the object, then the gripper was lowered. The "hit the stick" module (Module 4) lowers the gripper which under appropriate conditions result in that the ball jumps out of the gripper resulting in the goal state. Module 5 was created to handle stuck situations. This module lets the robot go backward and is applicable if the robot has not been able to move the wheels into the desired position for a while. This condition is a typical time-window based feature.

Simple case-analysis shows that there is no switching controller that would reach the goal with complete certainty (in the worst-case, the robot could re-

---

[5] Modules are numbered by the identification number of their features.

FNo	'on'	Behaviour
1	always	explore while avoiding obstacles
2	if the stick is in the viewing angle	go to the stick
3	if an object is near	examine the object grasp it if it is a ball
4	if the stick is near	hit the stick
5	if the robot is stuck	go backward
6	if the ball is grasped	-
7	if the stick is hit with the ball	-

Table 2. Description of the features and the modules. 'FNo.' means 'Feature No.', in the column labelled by 'on' the conditions under which the respective feature's value is 'on' are listed.

turn accidentally to state "10000000" from any state when the goal feature was 'off'), so that an almost-sure switching strategy should always exist. On the other hand, it is clear that a switching strategy which eventually attains the target does indeed exist.

3.3 Details of learning

A dense cost-structure was applied: the cost of using each behaviour was one except when the goal was reached, whose cost was set to zero. Costs were discounted at a rate of $\gamma = 0.99$. Note that from time to time the robot by chance became stuck (the robot's 'stuck feature' was 'on'), and the robot tried to execute a module which could not change the value of the feature-vector. This meant that the robot did not have a second option to try another module since by definition the robot could only make decisions if the feature-representation changed. As a result the robot could sometimes get stuck in a "perpetual" or so-called "jammed" state. To prevent this happening we built in an additional rule which was to stop and reinitialize the robot when it got stuck and could not unjam itself after 50 sensory measurements. A cost equivalent to the cost of never reaching the goal, i.e. a cost of $\frac{1}{1-\gamma}$ (= 100) was then communicated to the robot, which mimicked in effect that such actions virtually last forever.

Experiments were fully automated and organized in trials. Each trial run lasted until the robot reached the goal or the number of decisions exceeded 150 (a number that was determined experimentally), or until the robot became jammed. The 'stick was hit' event was registered by checking the state of the gripper (see also the description of **Feature 7**).

During learning the Boltzmann-exploration strategy was employed where the temperature was reduced by $T_{t+1} = 0.999\,T_t$ uniformly for all states [2]. During the experiments the cumulative number of successful trials were measured and compared to the total number of trials done so far, together with the average number of decisions made in a trial.

3.4 Results

Two sets of experiments were conducted. The first set was performed to check the validity of the module based approach, while the second was carried out to compare different RL algorithms. In the first set the starting exploration parameter T_0 was set to 100 and the experiment lasted for 100 trials. These values were chosen in such a way that the robot could learn a good switching policy, the results of these experiments being shown in Figure 2. One might

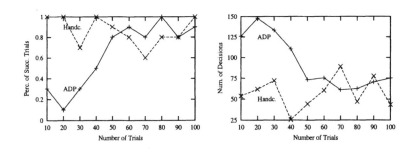

Fig. 2. Learning curves. In the first graph the percentage of successful trials out of ten are shown as a function of the number of trials. In the second graph the number of decisions taken by the robot and averaged over ten trials are both shown, as well as a function of the number of learning trials.

conclude from the left subgraph which shows the percentage of task completions in different stages of learning that the robot could solve the task after 50 trials fairly well. Late fluctuations were attributable to unsuccessful ball searches: as the robot could not see the balls if they were far from it, the robot had to explore to find one and the exploration sometimes took more than 150 decisions, yielding trials which were categorized as being failures. The evaluation of behaviour-coordination is also observed in the second subgraph, which shows the number of decisions per trial as a function of time. The reason for later fluctuations is again due to a ticklish ball search. The performance of a handcrafted switching policy is shown on the graphs as well. As can be seen the differences between the respective performances of the handcrafted and learnt switching functions are eventually negligible. In order to get a more precise evaluation of the differences the average number of steps to reach the goal were computed for both switchings over 300 trials, together with their standard deviations. The averages were 46.61 and 48.37 for the learnt and the handcrafted switching functions respectively, with nearly equal std-s of 34.78 and 34.82, respectively.

Theoretically, the total number of states is $2^7 = 128$, but as learning concentrates on feature-configurations that really occur this number transpires to be just 25 here. It was observed that the learnt policy was always consistent with a set of handcrafted rules, but in certain cases the learnt rules (which however can not be described here due to the lack of space) are more refined than their handcrafted counterparts. For example, the robot learnt to exploit the fact that the arena was not completely level and as a result balls were biased towards the stick and as a result if the robot did not hold a ball but could see the stick it moved towards the stick.

In the rest of the experiments we compared two versions of ARTDP and three versions of *real-time Q-learning* (RTQL). The two variants of ARTDP which we call ADP, and "ARTDP", corresponding to the cases when in the inner loop of ARTDP the optimal value function associated with the actual estimated model (transition probabilities and imediate cost) is computed and when only the estimate of the value of the actual state is updated. Note that due to the small number of states and module-based time discretization even ADP could be run in real-time. But variants of RTQL differ in the choice of the learning-rate's time-dependence. RTQL-1 refers to the choice of the so-called *search-then-converge* method, where $\alpha_k(s,a) = \frac{50}{100+n_k(s,a)}$, $n_k(s,a)$ being the number of times the event $(s,a) = (s_t,a_t)$ happened before time k plus one (the parameters 50 and 100 were determined experimentally as being the best choices). In the other two cases (the corresponding algorithms were denoted by RTQL-2 and RTQL-3 respectively) constant learning rates (0.1 and 0.25, respectively) were utilized.

The online performances of the algorithms were measured as the cumulative number of unsuccessful trials, i.e. the regret. The regret R_t at time t is the difference between the performance of an optimal agent (robot) and that of the learning agent accumulated up to trial t, i.e. it is the price of learning up to time t. A comparison of the different algorithms with different exploration ratios is given in Table 3. All algorithms were examined with all the four different exploration parameters since the same exploration rate may well result in different regrets for different algorithms, as was also confirmed in the experiments.

Method	$T_0 = 100$	$T_0 = 50$	$T_0 = 25$	$T_0 = 0$
ADP	(61;19;6)	(52;20;4)	(45;12;5)	(44;16;4)
ARTDP	36	29	50	24
RTQL-1	53	69	47	66
RTQL-2	71	65	63	73
RTQL-3	(83;1;2)	(79;26;3)	(65;24;4)	(61;14;2)

Table 3. Regret. The table shows the number of unsuccessful trials among the first 100 trials. The entries with three number in them show cases when more than one experiment was run. In those entries the first number shows the average of the number of unsuccessful trials, the second is the standard deviation while the third is the number of experiments run.

First note that in order to evaluate statistically the differences observed for different exploration strategies much more experiments would be needed but running these experiments would require an enormous amount of time (approximately 40 days) and have not been performed yet. Thus we performed the following procedure: Based on the first runs with every exploration-parameter and algorithm the algorithms that seemed to perform the best were selected (these were the ADP and the RTQL-3 algorithms) and some more experiments were carried out with these. The results of these experiments (15 more for the ADP and 7 more for the RTQL-3) indicated that the difference between the performances of the RTQL-3 and ADP is indeed significant at the level $p = 0.05$ (Student's t-test was applied for testing this).

We have also tested another exploration strategy which Thrun found the best among several undirected methods[6] [21]. These runs reinforced our previous findings that estimating a model (i.e. running ADP or ARTDP instead of Q-learning) could reduce the regret rate by as much as 40%.

4 Related Work

There are two main research-tracks that influenced our work. The first was the introduction of features in RL. Learning while using features were studied by Tsitsiklis and Van Roy to deal with large finite state spaces, and also to deal with infinite state spaces [22]. Issues of learning in partially observable environments have been discussed by Singh et al. [16].

The work of Connell and Mahadevan complements ours in that they set-up subtasks to be learned by RL and fixed the switching controller [13].

Asada et al. considered many aspects of mobile robot learning. They applied a vision-based state-estimation approach and defined "macro-actions" similar to our controllers [1]. In one of their papers they describe a goal-shooting problem in which a mobile robot shot a goal while avoiding another robot [24]. First, the robot learned two behaviours separately: the "shot" and "avoid" behaviours. Then the two behaviours were synthesised by a handcrafted rule and later this rule was refined via RL. The learnt action-values of the two behaviours were reused in the learning process while the combination of rules took place at the level of state variables.

Matarić considered a multi-robot learning task where each robot had the same set of behaviours and features [14]. Just as in our case, her goal was to learn a good switching function by RL. She considered the case when each of the robots learned separately and the ultimate goal was that learning should lead to a good collective behaviour, i.e. she concentrated mainly on the more involved multi-agent perspective of learning. In contrast to her work, we followed a more engineer-like approach when we suggested designing the modules based on well-articulated and simple principles and contrary to her findings it was discovered that RL can indeed work well at the modular level.

In the AI community there is an interesting approach to mobile robot control called Behaviour-Based Artificial Intelligence in which "competence" modules or behaviours have been proposed as the building blocks of "creatures" [12, 4]. The

---

[6] An exploration strategy is called undirected when the exploration does not depend on the number of visits to the state-action pairs.

decision-making procedure is, on the other hand, usually quite different from ours.

The technique proposed here was also motivated by our earlier experiences with a value-estimation based algorithm given in the form of "activation spreading" [20]. In this work activation spread out along the edges of a dynamically varying graph, where the nodes represented state transitions called triplets. Later the algorithm was extended so that useful subgoals could be found by learning [6, 7]. In the future we plan to extend the present module-based learning system with this kind of generalization capability. Such an extension may in turn allow the learning of a hierarchical organization of modules.

5 Summary and Conclusions

In this article module-based reinforcement learning was proposed to solve the co-ordination of multiple "behaviours" or controllers. The extended features served as the basis of time- and space discretization as well as the operating conditions of the modules. The construction principles of the modules were: decompose the problem into subtasks; for each subtask design controllers and the controllers' operating conditions; check if the problem could be solved by the controllers under the operating and observability conditions, add additional features or modules if necessary, set-up the reinforcement function and learn a switching function from experience.

The idea behind our approach was that a partially observable decision problem could be usually transformed into a completely observable one if appropriate features (filters) and controllers were employed. Of course some *a priori* knowledge of the task and robot is always required to find those features and controllers. It was argued that RL could work well even if the resulting problem was only an almost stationary Markovian. The design principles were applied to a real-life robot learning problem and several RL-algorithms were compared in practice. We found that estimating the model and solving the optimality equation at each step (which could be done owing to the economic, feature-based time-discretization) yielded the best results. The robot learned the task after 700 decisions, which usually took less than 15 minutes in real-time. We conjecture that using a rough initial model good initial solutions could be computed off-line which could further decrease the time required to learn the optimal solution for the task.

The main difference between earlier works and our approach here is that we have established principles for the design modules and found that our subsequent design and simple RL worked spendidly. Plans for future research include extending the method via module learning and also the theoretical investigation of almost stationary Markovian decision problems using the techniques developed in [10].

Acknowledgements

The authors would like to thank Zoltán Gábor for his efforts of building the experimental environment. This work was supported by the CSEM Switzerland, OTKA Grants No. F20132 and T017110, Soros Foundation No. 247/2/7011.

References

1. M. Asada, S. Noda, S. Tawaratsumida, and K. Hosoda. Purposive behavior acquisition for a real robot by vision-based reinforcement learning. *Machine Learning*, 23:279–303, 1996.
2. A. Barto, S. J. Bradtke, and S. Singh. Learning to act using real-time dynamic programming. *Artificial Intelligence*, 1(72):81–138, 1995.
3. R. Bellman. *Dynamic Programming*. Princeton University Press, Princeton, New Jersey, 1957.
4. R. Brooks. Elephants don't play chess. In *Designing Autonomous Agents*. Bradford-MIT Press, 1991.
5. T. Jaakkola, M. Jordan, and S. Singh. On the convergence of stochastic iterative dynamic programming algorithms. *Neural Computation*, 6(6):1185–1201, November 1994.
6. Z. Kalmár, C. Szepesvári, and A. Lőrincz. Generalization in an autonomous agent. In *Proc. of IEEE WCCI ICNN'94*, volume 3, pages 1815–1817, Orlando, Florida, June 1994. IEEE Inc.
7. Z. Kalmár, C. Szepesvári, and A. Lőrincz. Generalized dynamic concept model as a route to construct adaptive autonomous agents. *Neural Network World*, 5:353–360, 1995.
8. Z. Kalmár, C. Szepesvári, and A. Lőrincz. Module based reinforcement learning: Experiments with a real robot. *Machine Learning*, 31:55–85, 1998. joint special issue on "Learning Robots" with the J. of Autonomous Robots;.
9. M. Littman. *Algorithms for Sequential Decision Making*. PhD thesis, Department of Computer Science, Brown University, February 1996. Also Technical Report CS-96-09.
10. M. Littman and C. Szepesvári. A Generalized Reinforcement Learning Model: Convergence and applications. In *Int. Conf. on Machine Learning*, pages 310–318, 1996.
11. M. L. Littman, A. Cassandra, and L. P. Kaelbling. Learning policies for partially observable environments: Scaling up. In A. Prieditis and S. Russell, editors, *Proceedings of the Twelfth International Conference on Machine Learning*, pages 362–370, San Francisco, CA, 1995. Morgan Kaufmann.
12. P. Maes. A bottom-up mechanism for behavior selection in an artificial creature. In J. Meyer and S. Wilson, editors, *Proc. of the First International Conference on Simulation of Adaptive Behavior*. MIT Press, 1991.
13. S. Mahadevan and J. Connell. Automatic programming of behavior-based robots using reinforcement learning. *Artificial Intelligence*, 55:311–365, 1992.
14. M. Matarić. Reinforcement learning in the multi-robot domain. *Autonomous Robots*, 4, 1997.
15. S. Ross. *Applied Probability Models with Optimization Applications*. Holden Day, San Francisco, California, 1970.

16. S. Singh, T. Jaakkola, and M. Jordan. Learning without state-estimation in partially observable Markovian decision processes. In *Proc. of the Eleventh Machine Learning Conference*, pages pp. 284–292, 1995.
17. R. Sutton. *Temporal Credit Assignment in Reinforcement Learning*. PhD thesis, University of Massachusetts, Amherst, MA, 1984.
18. R. S. Sutton. Generalization in reinforcement learning: Successful examples using sparse coarse coding. *Advances in Neural Information Processing Systems*, 8, 1996.
19. C. Szepesvári and M. Littman. A unified analysis of value-function-based reinforcement-learning algorithms. *Neural Computation*, 1997. submitted.
20. C. Szepesvári and A. Lőrincz. Behavior of an adaptive self-organizing autonomous agent working with cues and competing concepts. *Adaptive Behavior*, 2(2):131–160, 1994.
21. S. Thrun. *The role of exploration in learning control*. Van Nostrand Rheinhold, Florence KY, 1992.
22. J. Tsitsiklis and B. Van Roy. An analysis of temporal difference learning with function approximation. Technical Report LIDS-P-2322, Laboratory for Information and Decision Systems, Massachusetts Institute of Technology, 1995.
23. J. N. Tsitsiklis and B. Van Roy. Feature-based methods for large scale dynamic programming. *Machine Learning*, 22:59–94, 1996.
24. E. Uchibe, M. Asada, and K. Hosoda. Behavior coordination for a mobile robot using modular reinforcement learning. In *Proc. of IEEE/RSJ Int. Conf. on Intelligent Robot and Sytems*, pages 1329–1336, 1996.
25. C. Watkins and P. Dayan. Q-learning. *Machine Learning*, 3(8):279–292, 1992.

Analysis and Design of Robot's Behavior: Towards a Methodology

Yassine Faihe and Jean-Pierre Müller

IIIA - University of Neuchâtel
Emile Argand 11
CH-2007 Neuchâtel - Switzerland.
Email: {yassine.faihe,jean-pierre.muller}@info.unine.ch

Abstract. We introduce a methodology to design reinforcement based control architectures for autonomous robots. It aims at systematizing the behavior analysis and the controller design. The methodology has to be seen as a conceptual framework in which a number of methods are to be defined. In this paper we use some more or less known methods to show the feasibility of the methodology. The postman-robot case study illustrates how the proposed methodology is applied.

1 Introduction

An autonomous robot is defined as a physical device which performs a predefined task in a dynamic and unknown environment without any external help. It has the ability to sense the state of the environment using sensors, to perform physical actions in the environment like object-grasping or locomotion, and has a control architecture which determines the action to be performed given a sensed state. Designing and then programming such an autonomous robot is very hard. The main reason is that we do not have to design an isolated robot but an interaction between this robot and an environment. Secondly, this interaction is stochastic and non-linear, and occurs through noisy sensors and actuators.

One way to overcome this difficulty is autonomous programming, that is make the robot acquire skills from the interaction with the environment. Such a process is called learning and refers to the ability to modify one's knowledge according to one's experience. Beside freeing the designer from explicitly programming the robot, learning is useful when the correct robot's behavior may change through time.

We are essentially interested in a particular learning paradigm called reinforcement learning and are investigating how it can be used to program autonomous robots. Reinforcement learning is concerned with the class of learning problems in which an agent operating in an environment learns how to achieve a given task by trial and error. To do so it adjusts its decision process on the basis of a form of feedback commonly called reinforcement signal. This feedback is a scalar which reflects an internal measure of the robot's performance. The robot is rewarded (punished) at each time step according to how close (how far) it is from the target behavior. Thus the optimal behavior would be the one maximising rewards and minimising punishments.

Reinforcement learning is very attractive because it raises the level of abstraction at which the designer intervenes (the robot is told what to do using the reinforcement function and not how to do it) and requires little programming effort (most of the work is done by autonomous training). Nevertheless and despite its mathematical foundations reinforcement learning can not be used as it is to make robots exhibit complex behaviors.

Such limitations are mainly due to the fact that both the complexity of the reinforcement function and the size of the state space with which the robot has to deal increase with the complexity of the behavior. Thus the learning process become slow and a wrong behavior may be generated. It is well known that breaking down the target behavior into simple behaviors which can work with simpler reinforcement functions and smaller state spaces, and then coordinating the component behaviors should give better results.

However, we claim that the big gap comes from the lack of understanding of the difference between a behavior and the mechanism which produces it. We argue that: (i) a behavior is the description from the observer's point of view, at different levels of abstraction, of a sequence of action produced by the robot via its coupling with the environment; (ii) complex behaviors may be produced by simple sensorimotor mechanisms interacting with the environment [1]; (iii) the design process of a behavior consists of a projection from the problem's domain (observer's point of view) to the co-domain (robot's point of view).

The programming process of an autonomous robot using reinforcement learning is not an easy job. A methodology which will guide the designer during the programming process is then needed.

Colombetti *et al.* [2] have developed a methodology and proposed a new technological discipline called Behavior Engineering. Their methodology called BAT (Behavior Analysis and Training) covers all the stages of the analysis and the design of an autonomous robots. The methodology assumes that the robot's apparatus and the environment are predefined, and the controller is endowed with a well-chosen learning system. The feasibility of this methodology has been demonstrated through three practical examples. However the decomposition and the reinforcement function specification rely heavily on the designer's intuition and past experience.

This article attempts to fill this gap and introduces a methodology which aims at systematizing the behavior analysis (behavior decomposition and reinforcement function specification) and the controller design. The proposed methodology has to be seen as a conceptual framework in which a number of methods are to be defined. In this paper we will use some more or less known methods to show the feasibility of the methodology.

The remainder of this paper is organized as follows. Section 3 describes a formal model of the interaction between the robot and its environment. Section 4 presents the proposed design methodology. Section 5 is devoted to a case study that illustrates the use of the methodology. Section 6 reports the experiments that have been carried out. Finally, the contribution of our methodology is discussed and suggestions for further improvements are provided.

2 The postman robot problem

2.1 The robot

The physical robot is a nomad 200 mobile platform. It has 16 infrared sensors for ranges less than 16 inches, 16 sonar sensors for ranges between 16 and 255 inches, and 20 tactile sensors to detect contact with objects. It is also equipped with wheel encoders and a compass to compute its current location and orientation relative to its initial ones. Finally, it has three wheels controlled together by two motors which make them translate and rotate. A third motor controls the turret rotation. These low level actuator movements constitute the *primitive commands* with which the robot acts in the world.

2.2 The task

The postman-robot is given a set of parallel and conflicting objectives and must satisfy them as best as it can. The robot acts in an office environment composed from offices, a battery charger and a mailbox. Its task is to collect letters from the offices and post them in the mailbox. While achieving its postman's task as efficiently as possible the robot has to avoid collisions with obstacles and recharge its batteries to prevent break-downs. The following assumptions are made:

- The robot can sense the number of letters it holds, its battery level and the number of letters in each office;
- The robot gets the letters once it is in an office, posts the letters once it is near the mailbox, and recharges its battery once it is near the charger (because the real robot we have does not have any grasping or recharging devices).

3 The model

In the proposed formal model (figure 1) the robot's behavior is modeled as a coupling of two dynamical systems: the environment and the robot (constituted by a sensorimotor loop).

3.1 The environment

In order to model the environment, we need to take into account its *state* including the robot itself (the robot being embedded and then part of the environment) and its *dynamics* under the influence of the robot action u. The law governing the environment's dynamics is a function which maps state-action pairs into next states: $x(t+1) = f(x(t), u(t))$. This law is often referred to as the *system state equation* or the *plant equation*. It is assumed that the environment's dynamics is both stochastic and stationary (i.e. f does not change over time).

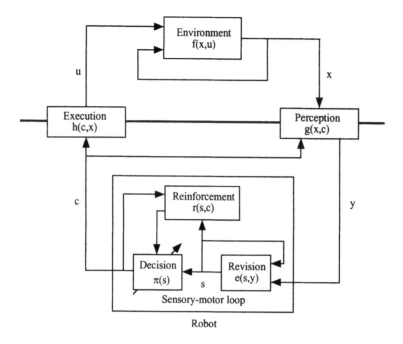

Fig. 1. Robot-environment interaction model

3.2 The robot

The robot is characterised by its perceptions (Y), its commands (C),as well as its decision and revision processes.

A clear distinction between the *environment state* $x \in X$ and the robot's *perception* $y \in Y$ is to be made. For example, the robot may have an obstacle in front of him and only gets a sensor reading. This measure is known by the observer as correlated with the distance to the obstacle, but *a priori* not by the robot.

In the same way, one must distinguish between the *command* $c \in C$ executed by the robot and the actual *action* $u \in U$ in the environment. For example, a robot may decide on a certain motor speed or number of wheel turns (the command) realising a movement in the environment (the action) which is not necessarily correlated and not perceived as such by the robot.

Now we have to introduce the notion of *internal state* $s \in S$ which is assumed Markovian and the revision process $e : S \times Y \to S$. The revision (or reconstruction) process can range from the identity function (when the perception is directly Markovian) up to the most sophisticated knowledge revision process.

The reinforcement function $r : S \times C \to \mathbb{R}$ is an internal measure of the robot performance. The robot is rewarded (punished) according to how close (how far) it is from the target behavior.

The decision process is modeled by a function called the *control policy* π : $S \to C$. The policy π is based on a utility function $Q(s, c)$ which gives a prediction of the return (i.e the expected discounted sum of reinforcements received) if the command c is executed in state s and the policy π is followed. The utility function Q can be computed through various reinforcement learning algorithms.

3.3 The coupling

The two dynamics mentioned above are coupled through the *perception* and *execution* devices.

The perception is a function of the environment's state (exteroception) and the robot's command (proprioception). It maps these stimuli to some internal representation. The perception is modeled as a function $g : X \times C \to Y$.

The distinction made between the command ($c \in C$) and the action in the environment ($u \in U$) becomes obvious when we notice that u has to be a function of both c and the current state x of the environment. Hence the execution of the same command in different states may produce different actions: the same number of wheel turns may result in different movement according to the space between the robot and an obstacle. Therefore the execution is modeled as a function $h : X \times C \to U$.

4 The methodology

The role of a methodology is to help the designer to develop quality products. Based on existing models and tools a methodology defines the goals to reach at each stage of the product's life cycle.

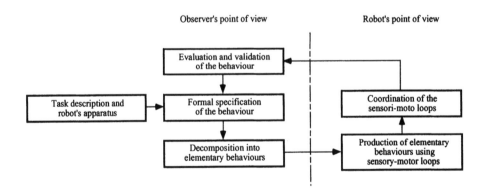

Fig. 2. Overview of the methodology

We will assume that both the robot's task and the necessary perceptions and primitive actions to accomplish are predefined. Primitives actions may contain some kind of servo control but remain the basic items out of which complex skills can be built. Given these assumptions and those above our methodology will show how:

- to formally specify the robot's behavior;
- to decompose the global behavior into a hierarchy of simple behaviors;
- to produce the elementary behaviors of the hierarchy using sensory-motor loops endowed with learning mechanisms;
- to coordinate the sensory-motor loops to get the behavior of the higher level;
- to evaluate and validate the global behavior.

In our definition of behavior we focused on two points of view : that of the external observer and that of the robot. While the former takes into account the environment's dynamics (including the robot because it is situated in it) the latter takes into consideration the robot's dynamics composed of the decision process. A third point of view is that of the designer which has to take into account the above dynamics as well as their interaction in order to come up with the suitable control architecture. Figure 2 gives an overview of the different stages of the methodology.

4.1   Specification and decomposition

In control theory one has to carefully formulate two aspects of the system.

The first of these is the system state equation. It describes the evolution law that the environment's state obeys for given actions. This is derived from the task statement and the description of the robot and the environment. The system state equation has to be specified at different level of abstraction starting from the highest one. The first question that the designer has to answer is what the robot can do, that is, what the robot commands are, in terms of high level behaviors. For example the postman-robot can *move toward the mailbox*, *move toward the batteries charger* and so on. The second question is what the relevant features x_i of the environment which change when one of the commands is performed are. These features become then the components of the state vector x. Finally the designer writes the state equations according to the environment's evolution. Even if a feature is directly measurable with sensors it is important to formulate its equation's evolution whenever possible, in order to show its link with other features which have not been found out before. This process is repeated until the command sets of the lowest level behaviors are only constituted by primitive commands.

The second aspect is the *performance criterion*. It is a function which evolve according to the trajectory of the system. The control policy is then to be chosen to optimize it. The performance criterion is a combination of an *objective function* and some *constraints* on the trajectory. The objective function $f(t)$ largely depends on the problem nature and represents an instantaneous measure of the system's performance such as the letters collected or the fuel consumption or the squared deviation from an optimal value at time t. The constraints set $C_s = \{x \in X \mid \varphi_1(x) = 0, ..., \varphi_n(x) = 0\}$ reflects the aspects of the trajectory of the process which are undesirable. Optimising the objective function while satisfying the constraints at the same time amounts to optimising an auxiliary function called *Hamiltonian function* defined as:

$$H(t, \lambda) = f(t) + \sum_i \lambda_i(x(t))\varphi_i(x(t))$$

for each state $x \in X$, where the Lagrange multipliers $\lambda_i(x) = 0$ if the constraints $\varphi_i(x) \leq 0$ is satisfied and $\lambda_i(x) = \lambda_{pi}$ otherwise. The positive constant λ_{pi} weights the strength of the penalty. This way to compute the Lagrange multipliers is called exterior penalties methods [10].

At the end of this stage a hierarchy of behaviors where each behavior is defined by a performance criterion is obtained. The lowest level behaviors directly interact with the environment using the primitive commands whereas behaviors of other levels use the learned behaviors of the level below.

4.2 Sensorimotor loop's design

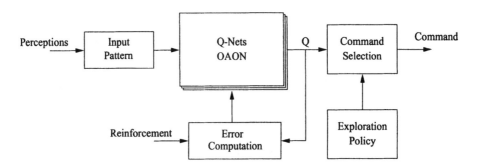

Fig. 3. The proposed generic sensorimotor loop.

In this section we shall present a generic sensorimotor loop which will allow us to generate a behavior given its specifications (state equations, objective function). The core of the sensorimotor loop is the learning system which is constituted by a set of neural networks. We have opted for neural networks because they are noise insensitive, are able to generalise over unseen situations, and can be used in a reccurent manner to deal with perceptual aliasing [6]. The neural networks are used to learn the Q-values by combining TD-learning techniques and back-propagation algorithm [13, 12]. To reduce interferences between actions we shall use one network per action (OAON architecture). The reinforcement function is directly computed from the objective function and has the form of a gradient: $r(t) = f(t) - f(t-1)$. When computed like that the reinforcement function gives a continuous information on the progress made by the robot and is not only a final reward when the goal is reached. Such a function can speed up the learning and allow better exploration [8]. Finally we need some exploration mechanism to guide the sensorimotor loop during its search. Thrun [14] has studied several methods ranging from the simple probability distribution to the more sophisticated active exploration mechanism.

4.3 Coordination

Generally stated the coordination problem within a hierarchy of sensory-motor loops consists in activating at each time step one sensory-motor loop at each level in order to generate the global expected behavior. A generic solution to this problem may consist in allocating an index to each sensory-motor loop and then activating the one with the highest index. The open question would then concern the meaning of such indexes and how to compute them.

The first idea that comes to mind is to treat the sensory-motor loops as macro-commands or closed-loop policies [11]. In this case the index will be the Q-value of this macro-command relative to the state of the upper sensorimotor loop that use it. Such a coordination has been used by Lin [7], and Dayan and Hinton [3], and recently generalised by Dietterich [4]. In this method the learning process is uniform through the levels of the hierarchy and the problem is solved at different levels of abstraction. However, from the second level of the hierarchy action selection is replaced by behavior selection and the time scale for decision making rises. This means that once a behavior is selected it will keep the control of the robot until its completion[1]. Hence learning will occur only at the termination of this behavior. Because a selected behavior can not be interrupted this methods may have some drawbacks in problems involving the satisfaction of multiple and concurrent objectives.

4.4 Evaluation and validation

In this stage we have to make the distinction between the behavior assessment [2] and the evaluation of the robot learning. The former is a qualitative criterion and the latter is a quantitative criterion. The behavior produced by the robot has to satisfy both of them in order to be valid.

To assess a behavior the designer should validate its *correctness* and its *robustness*. This is done from the observer's point of view. A behavior is correct when the task assigned to the robot is performed. For example we shall validate the postman-robot if we see the robot collecting and posting the letters without running out of energy. On the other hand a behavior is robust if it remains correct when structural changes of the environment occur.

From a quantitative point of view we can evaluate the asymptotic convergence to the optimal behavior [5] regarding to two criteria. The first criterion is the convergence speed. That is the necessary time (number of interaction cycles) to reach a plateau. The second criterion is the quality of the convergence. It is represented by the value of the reached plateau. Finally to evaluate the robot

---

[1] A behavior is completed when it reaches its goal or when it is not applicable anymore.

performances we should need some metrics. Cumulated deviations or average of received reinforcements over time are the most used metrics.

5 Case study: the postman robot

5.1 Specification and decomposition

The first level An analysis at the highest level of abstraction gives rise to the following commands:

u_c: go to the charger.
u_{mb}: go to the mailbox.
u_{oi}: go to office i (where $i = 1, \ldots, n$ and n is the number of offices).

These commands are exclusive and equal one if they are selected and zero otherwise. The state variables representing the relevant features of the environment at this level are:

x_r: number of letters that the robot holds.
x_b: robot battery level.
x_{li}: number of letters in office i.
x_{pc}: one if the robot is at the charger, zero otherwise.
x_{pmb}: one if the robot is at the mailbox, zero otherwise.
x_{poi}: one if the robot is in office i, zero otherwise.

The environment's dynamics equation are:

$$
\begin{aligned}
x_{li}(t+1) &= x_{li}(t) + \xi_i(t) - x_{poi}(t).(x_{li}(t) + \xi_i(t)) \\
x_r(t+1) &= x_r(t) + \sum_i x_{poi}(t).x_{li}(t) + x_{mb}.x_r(t) \\
x_{pc}(t+1) &= u_c(t) \\
x_{pmb}(t+1) &= u_{mb}(t) \\
x_{poi}(t+1) &= u_{oi}(t) \\
x_b(t+1) &= x_b(t) - k \sum_{i,j} d_{ij}.x_{pi}(t).x_{poi}(t+1) + x_{pc}.(x_{bmax} - x_b(t)),
\end{aligned}
$$

where d_{ij} is the distance between one place i and another place j (and vice versa) and k is a consumption factor. The letters flow in office i at time t is denoted by $\xi_i(t)$. The following initial conditions are used:

$$
\begin{aligned}
x_{li}(0) &= 0 \\
x_r(0) &= 0 \\
x_b(0) &= x_{bmax} \\
x_{p0}(0) &= 0 \\
x_{p1}(0) &= 1 \text{ (the robot may start at the mailbox).} \\
x_{pi}(0) &= 0
\end{aligned}
$$

The goal of the postman robot is to minimise the number of letters in the offices as well as the number of letters it holds by respectively collecting and then posting them. The following objective function is derived:

$$f_1(t) = \sum_i x_{li}(t) + \beta x_r(t), \qquad 0 < \beta < 1$$

subject to the battery level constraint:

$$\varphi(x) = x_b - x_{bth} \leq 0,$$

where x_{bth} is a safety threshold that depends upon the structure of the environment.

The letters carried by the robot may also be seen as a constraint and β as a Lagrange multiplier because the functions $\sum_i x_{li}(t)$ and $x_r(t)$ are antagonist: when the former is minimised the latter is maximised. Hence minimising $\sum_i x_{li}(t)$ and $x_r(t)$ amounts to minimising $\sum_i x_{li}(t)$ subject to $x_r(t) = 0$. The value of the Lagrange multiplier β is a constant between 0 and 1, so that any contribution to minimise either $\sum_i x_{li}(t)$ or $x_r(t)$ will also minimise $f_1(t)$. Moreover it is not necessary to set β to zero when the constraint is satisfied ($x_r(t) = 0$).

The performance criterion

$$H_1(t, \lambda) = \sum_i x_{li}(t) + \beta x_r(t) + \lambda(x(t)).\varphi(x(t))$$

is then deduced.

The second level At this level we have to specify the commands that will be used by the upper level. These commands can be seen as a generic behavior which consists in moving towards a goal. At each time step the robot execute a command which consists of a steering of Δ_θ degrees while moving at a constant speed. Thus the command set of the robot is constituted by several singletons $u_i = (\Delta_{\theta i})$.

The state variables are:

x_{si}: reading of sensor i. It indicates the distance to the nearest obstacle and depends on the spatial configuration of the environment as well as on the robot motions.

x_θ: robot orientation relative to the goal.

x_d: distance between the robot and the goal.

They evolve as follow:

$x_{si}(t + 1)$: this value is measurable.

$x_d(t + 1)$: this value is measurable using odometry.

$$x_\theta(t + 1) = arcos\left(\frac{x_d^2(t) - x_d^2(t+1) - \Delta_d^2}{2.x_d(t+1).\Delta_d}\right).sgn(x_\theta(t) + \Delta_\theta),$$

where $\Delta_d = x_d(t + 1) - x_d(t)$. The initial conditions are:

$$x_d(0) = d_{init},$$
$$x_\theta(0) = \theta_{init}.$$

To reach a goal the robot has to minimise its orientation relative to this goal while moving. This mean that the objective function

$$f_2(t) = x_\theta(t)$$

has to be minimised subject to the obstacles avoidance constraint

$$\varphi_i(x) = (x_{si} - d_{safe}) < 0,$$

where d_{safe} is the nearest safe distance to an obstacle. Like in the first level specification we obtain a Hamiltonian function as a performance criterion

$$H_2(t, \lambda) = x_\theta(t) + \sum_i \lambda_i(x(t))\varphi_i(x(t)).$$

Like in Millàn's work [9] it is assumed that the robot has a virtual sensor giving its distance to the goal. In further analysis this assumption will be discarded. The behavior *move toward a goal* may be decomposed into behaviors like *follow walls* and *pass a door*.

Through this specification and decomposition process the hierarchy of figure 4 is obtained.

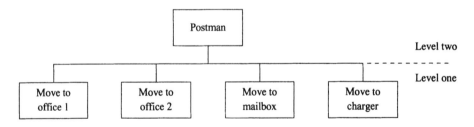

Fig. 4. The behaviors hierarchy of the postman robot.

5.2 Sensorimotor loop's design

We have used feed-forward neural network for the first level and a reccurent one for the second level. Both networks have a single output unit and a sigmoidal activation function. The input patterns are detailed in the experiments' section. A simple max-random probability distribution was used to provide probabilities for selecting commands: $P(a(t) = \arg\max_{a \in A} Q(x(t), a)|x(t))$. P is increased from 0 to 1 in N_{exp} exploration steps.

5.3 Coordination

The first level behaviors (navigation behaviors) are learned and used thereafter as macro-commands for the upper level. To do so we have implemented persistent neural networks.

5.4 Evaluation and validation

To judge the effectiveness of the overall behavior we defined the following metrics:

- the average letters in standby in the offices, the average letters carried by the robot as well as the average battery level for the external assessment;
- the average reward received for the learning process evaluation.

6 Experiments

6.1 Setups

The postman robot behavior was tested in an environment with two offices, one mailbox and one battery charger. Its size is approximately 13 m × 13 m. The experiments were carried out in the Nomad 200 simulator to which a letters flow simulator was added. For the second level the input pattern of the neural networks is a vector of 36 components. The sensors readings are formated to be in the interval [0, 1] and then connected to the first 16 components of the vector while the next 16 components are devoted to a coarse codification [9, 12] of the robot orientation relative to the goal. The remaining 4 components constitute the context and are connected to the hidden neurons. The input pattern of the first level neural networks is also a 32 bits vector. It is constituted of a binary codification of the number of letters in each offices as well as those transported by the robot and the battery level (4 × 7 bits). The robot location (in which office it is) is coded in the remaining 4 bits.

6.2 Results

The overall behavior was learned in two stages (modular learning). First the robot is trained to learn the low level behaviors (navigation) and then to learn the higher level behavior (task scheduling). After 30 trials the robot has learned how to move from one place to another. It takes him between 70 and 85 steps to reach any location from any starting point (the results of these experiments are not reported in this paper). Figures 5 and 6 show the obtained results for the second level behavior with a deterministic letters flow of 1 letter each 30 steps for office 1 and 1 letters each 40 steps for office 2. The performances of the learning robot are compared to those of the hand-coded robot endowed with the algorithm of figure 1.

With learning robot there are in average 5 letters in standby in the offices less than with the hand-coded robot whereas the average letters carried by robot rises by only 0.6 letters. Moreover the learning robot has a better energy management than the hand-coded one. The main reason of learning robot superiority is that it takes by itself and implicitly into account some parameters like the distance between the different places or the letters flow. The hand-coded algorithm will become complicated if these parameters are explicitly used.

Algorithm 1 Hand coded robot algorithm

---
if Battery-Level < Battery-Threshold **then**
  Move-To-Charger
else
  if Letters-Carried > (Letters-Office1 + Letters-Office2) **then**
    Move-To-Mailbox
  else
    if Letters-Office1 > Letters-Office2 **then**
      Move-To-Office1
    else
      if (Letters-Office1 = 0) and (Letters-Office2 = 0) **then**
        Move-To-Charger
      else
        Move-To-Office2
      end if
    end if
  end if
end if

---

7 Discussion and future work

We have proposed a methodology to design behavior based control architecture for autonomous robots. Through the postman robot problem we have shown how a complex behavior can be analysed and designed using the methodology as its associated methods. The following remarks are made:

- the controller's design is iterative. That is the results of the global behavior's evaluation can be used to correct the specifications. The cycle is repeated until the expected behavior is observed.
- the analysis process is top-down and from the observer's point of view while the design process is bottom-up and from the robot point of view;
- the distinction between the different points of view allows us to identify which parts have to be treated by the designer and which have to be learned by the robot. Hence we can easily combine engineering and evolution.

Our future work will be devoted to the development of efficient methods for the decomposition and coordination processes.

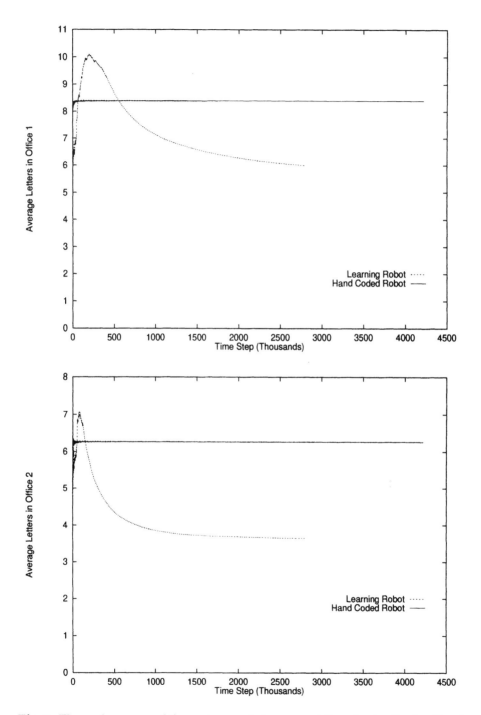

Fig. 5. The performances of the postman robot: average letters in standby in office 1 and office 2. A comparison between the learning and the hand-coded robots.

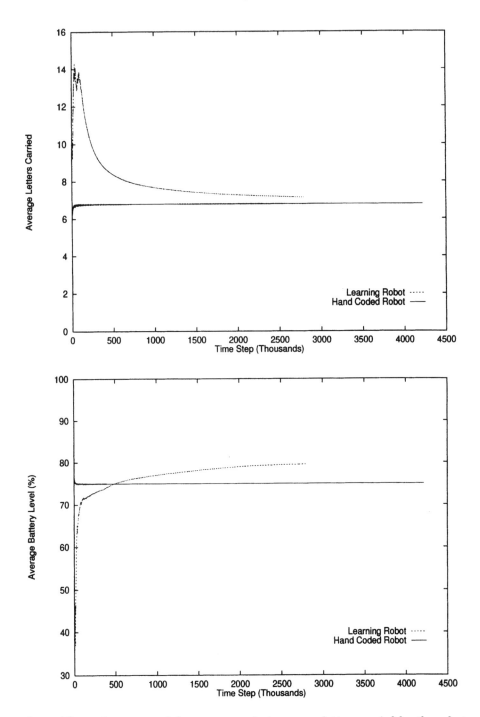

Fig. 6. The performances of the postman robot: average letters carried by the robot and average robot's battery level. A comparison between the learning and the hand-coded robots.

References

1. Valentino Braitenberg. *Vehicles. Experiments In Synthetic Psychology.* MIT Press, 1984.
2. Marco Colombetti, Marco Dorigo, and Giuseppe Borghi. Behavior analysis and design - a methodology for behavior engineering. *IEEE Transactions on Systems, Man and Cybernetics*, 26, 1996.
3. Peter Dayan and Geoffrey E. Hinton. Feudal reinforcement learning. In *Advances in Neural Information Processing Systems 5*, 1993.
4. Thomas G. Dietterich. Hierarchical reinforcement learning with the MAXQ value function decomposition. Technical report, Oregon State University, 1997.
5. Leslie P. Kaelbling, Michael L. Littman, and Andrew W. Moore. Reinforcement learning: a survey. *Journal of Artificial Intelligence Research*, 4, 1996.
6. Long J. Lin. Reinforcement learning with hidden state. In *Proceedings of the Second International Conference on Simulation of Adaptive Behavior*, 1992.
7. Long J. Lin. Hierarchical learning of robot skills by reinforcement. In *Proceedings of the IEEE International Conference on Neural Networks*, 1993.
8. Maja J. Mataric. Reward functions for accelerated learning. In W.W. Cohen and H. Hirsh, editors, *Proceedings of the Eleventh International Conference on Machine Learning*. Morgan Kaufmann, 1994.
9. José del R. Millàn. Rapid, safe and incremental learning of navigation strategies. *IEEE Transactions on Systems, Man and Cybernetics*, 26, 1996.
10. Michel Minoux. *Mathematical Programming*. John Wiley and Son, 1986.
11. Doina Precup, Richard S. Sutton, and Satinder P. Singh. Planning with closed-loop macro actions. Working notes of the 1997 AAAI Fall Symposium on Model-directed Autonomous Systems, 1997.
12. Gavin A. Rummery. *Problem Solving With Reinforcement Learning*. PhD thesis, University of Cambridge, 1995.
13. Richard S. Sutton. Implementation details of the TD(λ) procedure for the case of vector predictions and backpropagation. Technical Report TN87-509.1, GTE Laboratories, 1989.
14. Sebastian Thrun. Efficient exploration in reinforcement learning. Technical Report CMU-CS-92-102, Carnegie Mellon University, 1992.

Vision Based State Space Construction for Learning Mobile Robots in Multi-agent Environments

Eiji Uchibe[1], Minoru Asada[2] and Koh Hosoda[2]

[1] Dept. of Mechanical Engineering for Computer Controlled Machinery, Osaka University, Suita, Osaka 565-0871, Japan
[2] Dept. of Adaptive Machine Systems, Graduate School of Eng., Osaka University, Suita, Osaka 565-0871, Japan

Abstract. State space construction is one of the most fundamental issues for reinforcement learning methods to be applied to real robot tasks because they need a well-defined state space so that they can converge correctly. Especially in multi-agent environments, the problem becomes more difficult since visual information observed by a learning robot seems irrelevant to its self motion due to actions by other agents of which policies are unknown. This paper proposes a method which estimates the relationship between the learner's behaviors and the other agents' ones in the environment through interactions (observation and action) using the method of system identification to construct a state space in such an environment. In order to determine the state vectors of each agent, Akaike's Information Criterion is applied to the result of the system identification. Next, reinforcement learning based on the estimated state vectors is utilized to obtain the optimal behavior. The proposed method is applied to soccer playing physical agents, which learn to cope with a rolling ball and another moving agent. The computer simulations and the real experiments are shown and a discussion is given.

1 Introduction

Building a robot that learns to perform a task through visual information has been acknowledged as one of the major challenges facing Vision, Robotics and AI. Reinforcement learning has recently been receiving increased attention as a method for robot learning with little or no a priori knowledge and higher capability of reactive and adaptive behaviors [4].

In a multi-agent environment, the conventional reinforcement learning algorithm does not seem applicable because the environment including the other learning agents may change randomly from a viewpoint of the learning agent. There are two major reasons why the learning would be difficult in a multi-agent environment.

1. The other agent may use a stochastic action selector which could take a different action even if the same sensation of the learner occurs to it.

2. The other agent has a perception (sensation) different from the learning agent's. This means that the learning agent would not be able to discriminate different situations which the other agent can do, and vice versa.

Therefore, the learner cannot predict the other agent behaviors correctly even if its policy is fixed unless explicit communication is available. It is important for the learner to discriminate the strategies of the other agents and to predict their movements in advance to learn the behaviors successfully.

Littman [9] proposed the framework of Markov Games in which Q-learning agents try to learn a mixed strategy optimal against the worst possible opponent in a zero-sum 2-player game in a grid world. He assumed that the opponent's goal is given to the learner (opponent tries to minimize a single reward function, while it is to be maximized by the learning agent). Sandholm and Crites [11] studied the ability of a variety of Q-learning agents to play iterated prisoner's dilemma game against an unknown opponent. They showed that adequate previous moves and sensations are needed for the successful learning. Lin [8] compared recurrent-Q based on a recurrent network with window-Q based on both the current sensation and the N most recent sensations and actions, and he showed the former is superior to the latter because a recurrent network can cope with historical features appropriately. However, it is still difficult to determine the number of neurons and the structures of network in advance. Furthermore, their methods utilize the global information. Although the uncertainties of sensor and actuator outputs are considered by a stochastic transition model in the state space, such a model cannot account for the accumulation of sensor errors in estimating the robot position. Further, from the viewpoint of real robot applications, we should construct the state space so that it can reflect the outputs of the physical sensors which are currently available and can be mounted on the robot.

Robotic soccer is a good domain for studying multi-agent problems [6]. Stone and Veloso proposed *layered learning* method which consists of two levels of learned behaviors [12]. The lower is for basic skills such as interception of a moving ball and the higher is one which can make decisions whether or not to make a pass using decision tree. Uchibe et al. proposed a method of modular reinforcement learning which coordinates multiple behaviors taking account of a tradeoff between learning time and performance [13]. Since these methods utilize the current sensor outputs as states, their methods can not cope with temporal changes of image features caused by motions of objects.

As described above, the existing methods in multi agent environments need state vectors in order fort them to converge. However, it is difficult to obtain a reasonable analytical model in advance. Therefore, the modeling architecture is required to make the reinforcement learning applicable.

In this paper, we propose a method which estimates the relationship between the learner's behaviors and the other agents' through interactions (observation and action) using the method of system identification. In order to construct the local predictive model of other agents, we apply Akaike's Information Criterion(AIC) [1] to the result of Canonical Variate Analysis(CVA) [7], which is widely used in the field of system identification. The local predictive model is constructed based on the observation and action of the learner (observer).

We apply the proposed method to a simple soccer-like game in a physical environment. The task of the agent is to shoot a ball which is passed back from the other agent. Since the environment consists of the stationary agents (the goal and the line), a passive agent (the ball) and an active agent (the opponent), the learner has to construct the adequate models for these agents. After constructing the models and estimating their parameters, the reinforcement learning is applied in order to acquire purposive behaviors. The proposed method can cope with the moving ball because a state vector for learning is selected appropriately so as to predict the successive steps. The simulation results and the real experiments are shown and a discussion is given.

2 Construction of the internal state model from observation and action

2.1 Local predictive models for other agents

In order to succeed in learning, it is necessary for the learner to predict the successive situations as mentioned above. However, the agent can not obtain the complete information necessary to correctly predict them because of *partial observation* due to the limitation of sensing capability. Since we consider that the robot should construct the state space from its viewpoint, what the learning agent can do is to collect all the observed data with the motor commands taken during the observation and to estimate the relationship between the observed agents and the learner's behaviors in order to take an adequate behavior although it might not be guaranteed to be optimal. In the following, we consider to utilize a method of system identification, regarding the previous observed data and the motor commands as the inputs, and future observation results as the outputs of the system, respectively.

Figure 1 shows an overview of the proposed method consisting of local predictive models and reinforcement learning architecture. At first, the learning agent collects the sequence of sensor outputs and motor commands to construct the local predictive models, which are described in section 2.2. By approximating the relationship between inputs (learner's action) and outputs (observation), the local predictive model gives the learning agent not only the successive states of the agent but also the priority of state vectors, which means that first a few vectors might be sufficient to predict the successive states.

The flow of the proposed method is summarized as follows:

1. Collect the observation vectors and the motor commands (Section 2.2).
2. Estimate the state space with the full dimension directly from the observations and motor commands (Section 2.2).
3. Determine the dimension of the state vectors which is the result of the trade off between the error and the complexity of the model (Section 2.3).

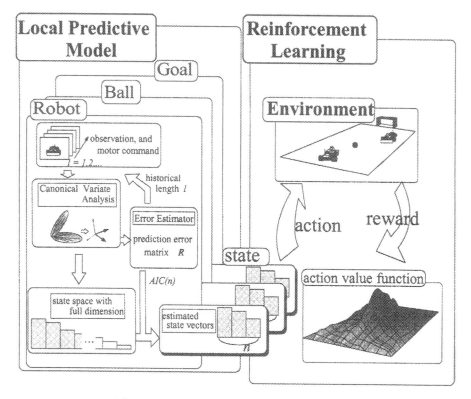

Fig. 1. An overview of the proposed method

4. Apply the reinforcement learning based on the estimated state vectors (Section 3).

2.2 Canonical Variate Analysis

A number of algorithms to identify multi-input multi-output (MIMO) combined deterministic-stochastic systems have been proposed [15]. In contrast to 'classical' algorithms such as PEM (Prediction Error Method), the subspace algorithms do not suffer from the problems caused by a priori parameterizations. Larimore's Canonical Variate Analysis (CVA) [7] is one of such algorithms, which uses canonical correlation analysis to construct a state estimator. We define $\boldsymbol{P}$ and $\boldsymbol{F}$ as follows : the past inputs and outputs

$$\boldsymbol{P} := \left(\boldsymbol{p}(1) \cdots \boldsymbol{p}(N/2 - 1) \right),$$

the future outputs

$$\boldsymbol{F} := \left(\boldsymbol{f}(N/2) \cdots \boldsymbol{f}(N) \right),$$

and we denote the future input block Hankel matrix as U. CVA algorithm is insensitive to scaling of the inputs (motor commands) and/or the outputs (sensor outputs), because the CVA algorithm considers only the angles and the normalized directions between the past inputs and outputs orthogonalized to the future inputs $(P/U^{\perp})$ and the future outputs orthogonalized to the future inputs $(F/U^{\perp})$ [15], where $A^{\perp}$ denotes the subspace perpendicular to the row space of A, and B/A is shorthand for the projection of the row space of B onto the row space of A (See Figure 2).

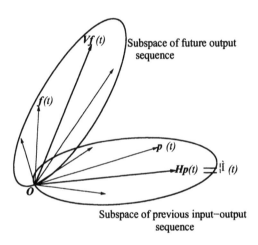

Subspace of future output sequence

Subspace of previous input–output sequence

Fig. 2. the subspace

Let $u(t) \in \Re^m$ and $y(t) \in \Re^q$ be the input and the output generated by the unknown system, respectively,

$$x(t+1) = Ax(t) + Bu(t) + w(t),$$
$$y(t) = Cx(t) + Du(t) + v(t), \tag{1}$$

with

$$E\left\{ \begin{bmatrix} w(t) \\ v(t) \end{bmatrix} \begin{bmatrix} w^T(\tau) & v^T(\tau) \end{bmatrix} \right\} = \begin{bmatrix} Q & S \\ S^T & R \end{bmatrix} \delta_{t\tau},$$

and $A, Q \in \Re^{n \times n}$, $B \in \Re^{n \times m}$, $C \in \Re^{q \times n}$, $D \in \Re^{q \times m}$, $S \in \Re^{n \times q}$, $R \in \Re^{q \times q}$. $E\{\cdot\}$ denotes the expected value operator and $\delta_{t\tau}$ the Kronecker delta. $v(t) \in \Re^q$ and $w(t) \in \Re^n$ are unobserved, Gaussian-distributed, zero-mean, white noise vector sequences. Larimore's Canonical Variate Analysis (CVA) [7] is one of identification algorithms, which uses canonical correlation analysis to construct a state estimator. CVA uses a new vector μ which is a linear combination of the previous input-output sequences since it is difficult to determine the dimension of $\boldsymbol{x}$. Eq.(1) is transformed as follows:

$$\begin{bmatrix} \boldsymbol{\mu}(t+1) \\ \boldsymbol{y}(t) \end{bmatrix} = \boldsymbol{\Theta} \begin{bmatrix} \boldsymbol{\mu}(t) \\ \boldsymbol{u}(t) \end{bmatrix} + \begin{bmatrix} \boldsymbol{T}^{-1}\boldsymbol{w}(t) \\ \boldsymbol{v}(t), \end{bmatrix}, \tag{2}$$

where

$$\hat{\boldsymbol{\Theta}} = \begin{bmatrix} \boldsymbol{T}^{-1}\boldsymbol{A}\boldsymbol{T} & \boldsymbol{T}^{-1}\boldsymbol{B} \\ \boldsymbol{C}\boldsymbol{T} & \boldsymbol{D} \end{bmatrix}, \tag{3}$$

where $\boldsymbol{x}(t) = \boldsymbol{T}\boldsymbol{\mu}(t)$. We follow the simple explanation of the CVA method.

1. For $\{\boldsymbol{u}(t), \boldsymbol{y}(t)\}$, $t = 1, \cdots N$, construct new vectors

$$\boldsymbol{p}(t) = \begin{bmatrix} \boldsymbol{u}(t-1) \\ \vdots \\ \boldsymbol{u}(t-l) \\ \boldsymbol{y}(t-1) \\ \vdots \\ \boldsymbol{y}(t-l) \end{bmatrix}, \quad \boldsymbol{f}(t) = \begin{bmatrix} \boldsymbol{y}(t) \\ \boldsymbol{y}(t+1) \\ \vdots \\ \boldsymbol{y}(t+k-1) \end{bmatrix}.$$

2. Compute the estimated covariance matrices $\hat{\boldsymbol{\Sigma}}_{pp}$, $\hat{\boldsymbol{\Sigma}}_{pf}$ and $\hat{\boldsymbol{\Sigma}}_{ff}$, where $\hat{\boldsymbol{\Sigma}}_{pp}$ and $\hat{\boldsymbol{\Sigma}}_{ff}$ are regular matrices.
3. Apply singular value decomposition

$$\hat{\boldsymbol{\Sigma}}_{pp}^{-1/2} \hat{\boldsymbol{\Sigma}}_{pf} \hat{\boldsymbol{\Sigma}}_{ff}^{-1/2} = \boldsymbol{U}_{aux}\boldsymbol{S}_{aux}\boldsymbol{V}_{aux}^T, \tag{4}$$
$$\boldsymbol{U}_{aux}\boldsymbol{U}_{aux}^T = \boldsymbol{I}_{l(m+q)}, \quad \boldsymbol{V}_{aux}\boldsymbol{V}_{aux}^T = \boldsymbol{I}_{kq},$$

and $\boldsymbol{U}$ and $\boldsymbol{V}$ are calculated as:

$$\boldsymbol{U} := \boldsymbol{U}_{aux}^T \hat{\boldsymbol{\Sigma}}_{pp}^{-1/2},$$
$$\boldsymbol{V} := \boldsymbol{V}_{aux}^T \hat{\boldsymbol{\Sigma}}_{ff}^{-1/2}.$$

4. The n dimensional new vector $\boldsymbol{\mu}(t)$ is defined as:

$$\boldsymbol{\mu}(t) = [\boldsymbol{I}_n \ 0]\boldsymbol{U}\boldsymbol{p}(t), \tag{5}$$

5. Estimate the parameter matrix $\boldsymbol{\Theta}$ by applying the least square method to Eq (2).

Strictly speaking, the learning agent should construct one local predictive model about the whole system since all the agents do in fact interact. However, it is intractable to collect the adequate input-output sequences and estimate the proper model because the dimension of state vector drastically increases. Therefore, the learning (observing) agent obtains the local predictive models by applying the CVA method to all the (observed) agents separately.

2.3 Determination of the dimension of other agent

It is important to decide the dimension n of the state vector $\boldsymbol{\mu}$ and lag operator l that tells how long the historical information is involved in determining the size of the state vector when we apply CVA to the classification of agents. Although the estimation is improved if l is larger and larger, much more historical information is necessary. However, it is desirable that l is as small as possible with respect to the memory size. For n, complex behaviors of other agents can be captured by choosing the order n high enough, but we have to take account of the trade off between the number of parameters and the precision of the estimation.

In order to determine n, we apply Akaike's Information Criterion (AIC) which is widely used in the field of time series analysis. AIC is a method for balancing precision and computation (the number of parameters). Let the prediction error be ε and covariance matrix of error be

$$\hat{\boldsymbol{R}} = \frac{1}{N - k - l + 1} \sum_{t=l+1}^{N-k+1} \varepsilon(t)\varepsilon^T(t).$$

Then $AIC(n)$ is calculated by

$$AIC(n) = (N - k - l + 1)\log|\hat{\boldsymbol{R}}| + 2\lambda(n), \qquad (6)$$

where

$$\lambda(n) = n(2p + m) + pm + \frac{1}{2}p(p + 1). \qquad (7)$$

The optimal dimension n^* is defined as

$$n^* = \arg\min_n AIC(n),$$

where

$$1 \leq n \leq \min(l(m + q), kq).$$

However, the parameter l is not under the influence of the $AIC(n)$. Because the reinforcement learning algorithm is applied to the result of the estimated state vector in order to cope with the non-linearity and the error of modeling, the learning agent does not have to construct the *strict* local predict model. Therefore, we utilize $\log|\hat{\boldsymbol{R}}|$ to determine l.

1. Memorize the q dimensional vector $\boldsymbol{y}(t)$ about the agent and m dimensional vector $\boldsymbol{u}(t)$ as a motor command.
2. From $l = 1\cdots$, identify the obtained data.
   (a) If $\log|\hat{\boldsymbol{R}}| < 0$, stop the procedure and determine n based on $AIC(n)$,
   (b) else, increment l until the condition (a) is satisfied or $AIC(n)$ does not decrease.

3 Reinforcement Learning

After estimating the state space model represented by Eq. 2, the agent begins to learn behaviors using a reinforcement learning method. Q learning [16] is a form of model-free reinforcement learning based on the stochastic dynamic programming. It provides robots with the capability of learning to act optimally in a Markovian environment. In the previous section, the appropriate dimension n of the state vector $\boldsymbol{\mu}(t)$ has been determined, and the successive state can be predicted. Therefore, we regard the environment as a Markovian one.

In order to utilize the result of identification for the Q learning, the state vector $\boldsymbol{\mu}$ has to be quantized. Because the state vector $\boldsymbol{\mu}$ is calculated as

$$E\{\boldsymbol{\mu\mu}^T\} = \boldsymbol{I}_n,$$

we segment the state space as

$$\mu_i < -1, \ -1 \leq \mu_i < 1, 1 \leq \mu_i, \quad \text{for all } i.$$

Hereafter, we denote the estimated state vector $\boldsymbol{\mu}$ as $\boldsymbol{x}$ for reader's understanding. We assume that the robot can discriminate the set $\boldsymbol{X}$ of distinct world states, and can take the set $\boldsymbol{U}$ of actions on the world. A simple version of a Q learning algorithm used here is shown as follows.

1. Initialize $Q(x, u)$ to 0s for all combination of $\boldsymbol{X}$ and $\boldsymbol{U}$.
2. Perceive current state x.
3. Choose an action u according to action value function.
4. Carry out action u in the environment. Let the next state be x' and immediate reward be r.
5. Update action value function from x, u, x', and r,

$$Q_{t+1}(x, u) = (1 - \alpha_t)Q_t(x, u) + \alpha_t(r + \gamma \max_{u' \in \boldsymbol{U}} Q_t(x', u')) \tag{8}$$

   where α_t is a learning rate parameter and γ is a fixed discounting factor between 0 and 1.
6. Return to 2.

4 Task and Assumptions

We apply the proposed method to a simple soccer-like game including two active agents (See Figure 3). Each agent has a single color TV camera and does not know the locations, the sizes and the weights of the ball and the other agent, any camera parameters such as focal length and tilt angle, or kinematics/dynamics of itself. They move around using a 4-wheel steering system. The effects of an action against the environment can be informed to the agent only through the visual information. As motor commands, each agent has 7 actions such as go straight,

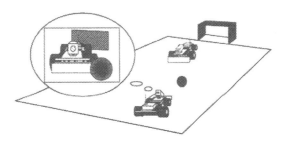

Fig. 3. The environment and our mobile robot

turn right, turn left, stop, and go backward. Then, the input $\boldsymbol{u}$ is defined as the 2 dimensional vector as

$$\boldsymbol{u}^T = [v \ \phi], \quad v, \phi \in \{-1, 0, 1\},$$

where v and ϕ are the velocity of motor and the angle of steering respectively and both of which are quantized.

The output (observed) vectors are shown in Figure 4. In case of the ball,

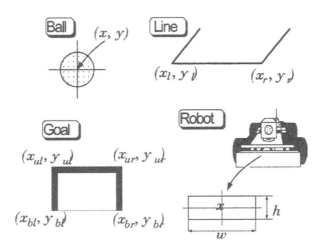

Fig. 4. Image feature points of the ball, goal, line and agent

the center position of the ball image (x_c, y_c) is used, and the both ends (x_l, y_l) and (x_r, y_r) are used for the field lines. In case of the goal, the four corners of the goal image (x_{ul}, y_{ul}), (x_{bl}, y_{bl}), (x_{ur}, y_{ur}), and (x_{br}, y_{br}) are used. In case of other agent, the center of position, the width and the height of the front plate are used. As a result, the dimension of the observed vector for the ball, the goal, the line, and the agent are 2, 4, 8, and 3 respectively.

We assume that the other active agent has some basic behaviors designed by programmer such that 1) to move randomly, or 2) to move to the ball, and that it does not change its behavior frequently.

5   Experimental Results

5.1   Simulation Results

Table 1. The estimated dimensions in computer simulation

agent	l	n	$\log |R|$	AIC
line	1	3	−2.14	−800
goal	1	2	−0.001	121
ball	2	4	0.232	138
random walk	3	6	1.22	232
move to the ball	3	6	−0.463	79

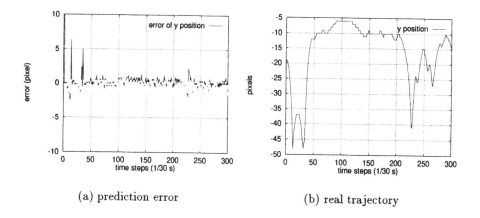

(a) prediction error   (b) real trajectory

Fig. 5. Prediction errors of the y position of the ball

Table 1 shows the result of identification. In order to predict the successive situations $l = 1$ is sufficient for the goal and line, while the ball needs 2 steps.

Figures 5 show the result (error and trajectory) of the ball. At time steps 40 and 230, the learner kicked the ball, therefore prediction errors become large drastically.

The observing agent can not predict the random walk agent as a matter of course. The agent which moves to the ball can be identified by the same dimension of the random agent, but the prediction error is much smaller than that of the random walk.

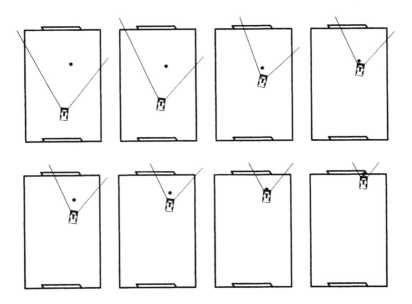

Fig. 6. The robot succeeded in shooting a moving ball into a goal

Figure 6 shows a sequence of shooting a slowly moving ball into the goal using the proposed method. We assign a reward value 1 when the ball was kicked into the goal or 0 otherwise and the environment consists of the ball, the goal and the line. The two lines emerged from the agent show its visual angle. If the learning agent uses the only current observation as the state vectors about the ball and the goal, the leaning agent can not acquire the optimal behavior when the ball is rolling. In other words, the action value function does not become to be stable because the state and action spaces are not consistent with each other.

Next, we show a sequence of passing a ball toward the other agent in Figure 7. Passing a ball to the other agent is regarded as shooting (kicking) a ball toward the moving goal. We assign a reward value 1 when the ball was kicked into the other agent, −0.8 when the learner makes a collision with the other agent or 0 otherwise and the environment consists of the ball and the other agent. Table

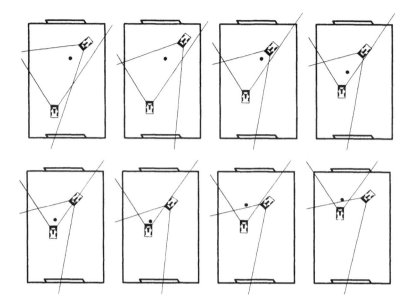

Fig. 7. The robot succeeded in passing a ball toward the other agent

2 shows a comparison about the success rate of shooting and passing with the model based on only the current perception and action.

Table 2. Comparison between the proposed method and the other one using only the current observation

state vector	success of shooting (%)	success of passing (%)
current sensor space	10.2	9.8
estimated state space by the proposed method	78.5	53.2

5.2 Real Experiments

We have constructed the radio control system of the robot, following the remote-brain project by Inaba et al. [5]. Figure 8 shows a configuration of the real mobile robot system. The image taken by a TV camera mounted on each robot is transmitted to a UHF receiver and processed by Datacube MaxVideo

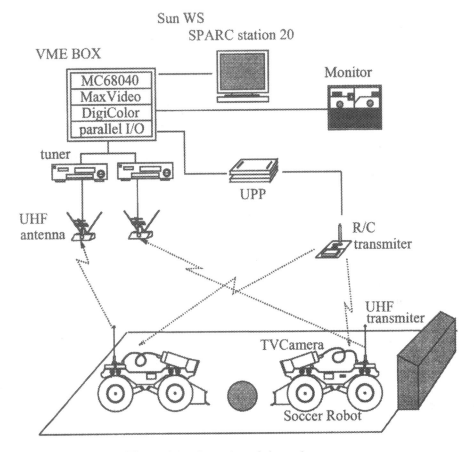

Fig. 8. A configuration of the real system.

200, a real-time pipeline video image processor. In order to simplify and speed up the image processing time, we painted the ball, the goal, and the opponent in red, blue, and yellow, respectively. The input NTSC color video signal is first converted into HSV color components in order to easily extract the objects. Figure 9 shows the images taken by a TV camera mounted on each robot (left : shooter, and right : passer). The image processing and the vehicle control system are operated by VxWorks OS on MC68040 CPU which are connected with host Sun workstations via Ether net. The tilt angle is about -26 [deg] so that robot can see the environment effectively. The horizontal and vertical visual angles are about 67 [deg] and 60 [deg], respectively.

The task of the passer is to pass a ball to the shooter while the task of the shooter is to shoot a ball into the goal. Table 3 and Figure 10 show the experimental results. The value of l for the ball and the agent are bigger than that of computer simulation because of the noise of the image processing and the

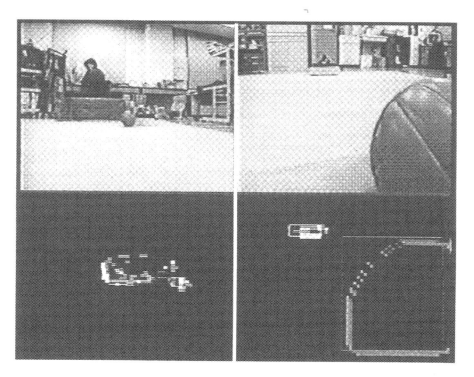

Fig. 9. input images and processed images (left : shooter, and right : passer)

Table 3. The estimated dimensions in the real environment

	from the shooter				from the passer			
	l	n	$\log\lvert R\rvert$	AIC	l	n	$\log\lvert R\rvert$	AIC
ball	4	4	1.88	284	4	4	1.36	173
goal	1	3	−1.73	−817				
robot	5	4	3.43	329	5	4	2.17	284

dynamics of the environment such as the eccentricity of the centroid of the ball. Even though the local predictive model of the same ball for each agent is similar ($n = 4$, and slight difference in $\log\lvert R\rvert$ and AIC) (See Table3), the estimated state vectors are different from each other because there are differences in several factors such as tilt angle, the velocity of the motor and the angle of steering. We checked what happened if we replace the local predictive models between the passer and the shooter. Eventually, the large prediction errors of both side were observed. Therefore the local predictive models can not be replaced between physical agents. Finally, Figure 11 shows a sequence of images where the passer kicked a ball towards the shooter, which shot it into the goal.

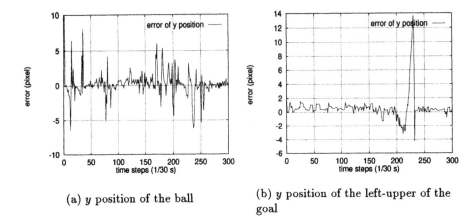

(a) y position of the ball

(b) y position of the left-upper of the goal

Fig. 10. Prediction error in real environment

6 Conclusion

This paper presents a method of behavior acquisition in order to apply reinforcement learning to the environment including other agents. Our method takes account of the trade-off among the precision of prediction, the dimension of state vector, and the length of steps to identify the model. Our robots can shoot and pass a ball even if a ball is rolling well.

As future work we plan to challenge the following issues:

- The local predictive model provided the state vectors by which prediction can be effectively done because they have strongly correlation between the past inputs/outputs and the future outputs. In order to accomplish the more complicated task, the learning robot can determine the minimum dimension of the state vectors in accordance with the increase of the level of the task complexity [14]. We are planning to extend the "Learning from Easy Missions" paradigm [3] to the complicated task.
- In our experiments, we quantized each of the elements of the estimated state vectors into three categories based on its variance. Several segmentation methods such as Parti game algorithm [10] and Asada's method [2] can be alternative.
- In the current system, we consider just two robots, and regard that the current system can cope with global interactions by reinforcement learning. However, more robots in the field we have, more complicated and higher interactions occur. Therefore, we challenge to extend our method when more than two robots learn cooperative and competitive behaviors.

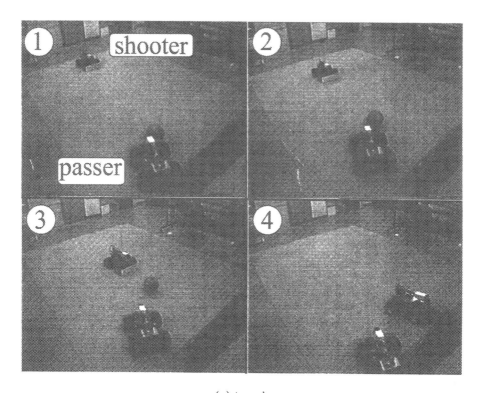

(a) top view

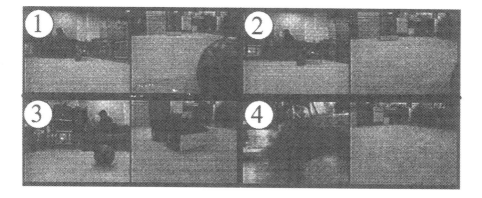

(b) obtained images (left:shooter, right:passer)

Fig. 11. Acquired behavior

References

1. H. Akaike. A new look on the statistical model identification. *IEEE Trans. AC-19*, pages 716–723, 1974.
2. M. Asada, S. Noda, and K. Hosoda. Action-based sensor space categorization for robot learning. In *Proc. of the 1996 IEEE/RSJ International Conference on Intelligent Robots and Systems*, 1996.
3. M. Asada, S. Noda, S. Tawaratsumida, and K. Hosoda. Vision-based reinforcement learning for purposive behavior acquisition. In *Proc. of IEEE International Conference on Robotics and Automation*, pages 146–153, 1995.
4. J. H. Connel and S. Mahadevan. *Robot Learning*. Kluwer Academic Publishers, 1993.
5. M. Inaba. Remote-brained robotics : Interfacing AI with real world behaviors. In *Preprints of ISRR'93*, Pitsuburg, 1993.
6. H. Kitano, M. Asada, Y. Kuniyoshi, I. Noda, E. Osawa, and H. Matsubara. Robocup a challenge problem for ai. *AI Magazine*, 18(1):73–85, 1997.
7. W. E. Larimore. Canonical variate analysis in identification, filtering, and adaptive control. In *Proc. 29th IEEE Conference on Decision and Control*, pages 596–604, Honolulu, Hawaii, December 1990.
8. L.-J. Lin and T. M. Mitchell. Reinforcement learning with hidden states. In *Proc. of the 2nd International Conference on Simulation of Adaptive Behavior: From Animals to Animats 2.*, pages 271–280, 1992.
9. M. L. Littman. Markov games as a framework for multi-agent reinforcement learning. In *Proc. of the 11th International Conference on Machine Learning*, pages 157–163, 1994.
10. A. W. Moore and C. G. Atkeson. The parti-game algorithm for variable resolution reinforcement learning in multidimensional state-spaces. *Machine Learning*, 21:199–233, 1995.
11. T. W. Sandholm and R. H. Crites. On multiagent Q-learning in a semi-competitive domain. In *Workshop Notes of Adaptation and Learning in Multiagent Systems Workshop, IJCAI-95*, 1995.
12. P. Stone and M. Veloso. Using machine learning in the soccer server. In *Proc. of IROS-96 Workshop on Robocup*, 1996.
13. E. Uchibe, M. Asada, and K. Hosoda. Behavior coordination for a mobile robot using modular reinforcement learning. In *Proc. of the 1996 IEEE/RSJ International Conference on Intelligent Robots and Systems*, pages 1329–1336, 1996.
14. E. Uchibe, M. Asada, and K. Hosoda. Environmental complexity control for vision-based learning mobile robot. In *Proc. of IEEE International Conference on Robotics and Automation*, 1998 (to appear).
15. P. Van Overschee and B. De Moor. A unifying theorem for three subspace system identification algorithms. *Automatica*, 31(12):1853–1864, 1995.
16. C. J. C. H. Watkins and P. Dayan. Technical note: Q-learning. *Machine Learning*, pages 279–292, 1992.

Transmitting Communication Skills Through Imitation in Autonomous Robots

Aude Billard and Gillian Hayes

University of Edinburgh, Dept. of Artificial Intelligence
5 Forrest Hill, EH1 2QL Edinburgh
E-mail: audeb@dai.ed.ac.uk

Abstract. Sharing a common context of perception is a prerequisite in order for several agents to develop a common understanding of a language. We propose a method, based on a simple imitative strategy, for transmitting a vocabulary from a teacher agent to a learner agent. A learner robot follows and thus implicitly imitates the movements of a teacher robot. While bounded by mutual following, learner and teacher agents are set in a position from which they share a common set of perceptions, that they can then correlate with an arbitrary set of signals, a vocabulary. The teacher robot leads the learner robot through a series of situations in which it teaches it a vocabulary to describe its observations. The learner robot *grounds* the teacher's 'words' by associating them with its own perceptions. Learning is provided by a dynamical recurrent associative memory, an Artificial Neural Network architecture. The system is studied through simulations and physical experiments where the vocabulary concerns the agents' movements and orientation. Experiments are successful, learning is fast and stable in the face of a significant amount of experimental noise. This work suggests then that a simple movement imitation strategy is an interesting scenario for the transmission of a language, as it is an easy means of getting the agents to share the same physical context.

1 Introduction

Communication is a social skill and as such would be desirable for autonomous agents that are expected to interact with other agents or humans ([5], [6]). In particular, collaborative tasks would be easier if the agents could inform each other of what they are doing ([8]). For communication to be successful, the agents must share a common interpretation of the language they use. Whilst a fixed protocol could be designed to this end, we expect the system to be more robust and adaptable if the agents would develop their understanding of the language through a learning process. This work studies a method for transmitting a language from a teacher agent to a learner agent.

Recent simulation studies of language acquisition tend to address essentially the question of the origin of language and how it can evolve among a population (e.g. [9], [12], [1], [10]). In our approach, we focus on the learning and understanding of a language at the level of the individual rather than at the level of the society. We propose a method, in terms of the agent's behavioural capacities, whereby a meaning can be associated with an abstract signal, i.e. a 'word'. To our knowledge, this physical approach to the problem of the language understanding has very seldom been questioned nor investigated.

Sharing a common context, and so a common set of perceptions, is a prerequisite for the development of a common understanding of a language by different agents. In order for the learner agent to successfully interpret the teacher's words, it has to be able to make the same observations to which the teacher is referring. Sharing the same spatial and temporal context of perceptions is often not enough and attentional mechanisms must be used to restrict the set of perceptions to a relevant subset shared by both agents. In related robotics studies, this reduction of the information was enabled by pointing to the referred-to objects [11] or by using global knowledge of the learner's perception [14]. These methods are costly in terms of pre-programming and not easily adaptable to different agents or tasks. In this work, we use a simple imitative method, namely mutual following of teacher and learner agents. While bounded by mutual following, learner and teacher agents face the same direction and thus share a similar if not identical (due to different sensor sensitivity) set of external perceptions. This set of perceptions forms a subset of all the possible observations the agents could make from the same location. In addition, while following the teacher, the learner agent replicates the teacher's movements and thus shares similar internal perception of movement and energy consumption, while restricting itself to a subset of possible actions.

Learning to communicate means in our experiments that one agent, the learner tries to achieve a similar 'interpretation' of a situation to that of a second agent, the teacher. While followed by the learner, the teacher robot leads the learner robot through a series of situations in which it teaches it a vocabulary to describe its observations. In this context, the 'language' consists of a simple vocabulary, where the 'words' are (radio) signals that label specific configurations of sensor-actuator state. The teacher sends distinct signals to discriminate between different movements and orientations. The learner robot *grounds* the teacher's 'words' by associating them with its own perceptions, i.e. sensor measurements and motor states.

Other works used movement imitation for teaching a robot about movement related tasks ([4], [7]). Here, we propose to use this method for teaching a language. Movement imitation, as a means to learning a language, is interesting compared to a conventional supervised learning method, as used e.g. in [14], for at least two reasons: Firstly, it enables the learning agent to share a perception of the world similar to the teacher's. Secondly, it enables the teacher to direct the learner's learning. Teaching can concern both proprio and exteroreceptive perceptions. By leading the learner to specific locations, the teacher can teach it a vocabulary about objects or situations. By directing the learner's movements, the teacher can teach it a vocabulary about its internal states (position, movements or energy consumption). In the following strategy we propose, movement imitation is only implicit and not the result of a particular cognitive mechanism.

The rest of this paper is divided as follows. Section 2 describes the controller architecture of teacher and learner agents. Section 3 presents the experimental procedure of the simulation studies and physical experiments. Section 4 reports and discusses the results of these experiments, on which we conclude this paper in section 5.

2 Controller Architecture

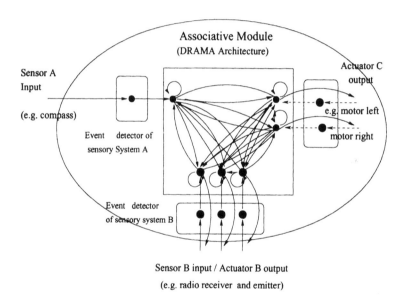

Fig. 1. Schematic representation of the control system of the learner and teacher agents with three sensor-actuator systems, e.g. compass, radio and motors. Learning is dynamical and performed continuously during the experiment. At each processing cycle, the learner robot's sensor-actuator information is compared to the measurements collected the previous cycle. Whenever a variation in the measurement is detected (by the *event recognition* module), the new measurement is presented to the associative architecture (DRAMA) to be associated with all simultaneous and previously recorded measurements in the same or other sensor-actuator system. DRAMA is a fully recurrent Neural Network. Its recurrent structure provides a short term memory of the measurements. Long term memory is obtained by updating the internal connections following Hebbian rules.

Learning to communicate is implemented in our experiments as part of a general multiple sensory association process, produced by a Dynamic Recurrent Associative Memory Architecture (DRAMA), previously described in [2]. We here briefly summarise its main properties, together with a description of the global control system of the agent:

1) It combines two connectionist model types, associative memory (Willshaw type) and recurrent networks, in one single network structure for achieving learning of temporally coincident patterns. 2) The network is fully connected with self-recurrent connections on each node. 3) Each connection in the network is associated with a time parameter and a confidence factor. The time parameters represent the temporal correlation between events while the confidence factors keep a memory of the frequency of a pattern occurrence. 4) Time parameters and confidence factors are updated following Hebbian rules, providing an associative type of learning. The closer the temporal occurrence of the events, the stronger the correlation. 5) Encoding of inputs and outputs is binary. 6) The self-recurrent connections on the units provide a short-term memory of one event per unit, whose maximal duration is fixed. 7) Long term memory of sequential and simultaneous correlations is provided by updating the connections between different units. 8) Data retrieving depends on the value of the time parameters and confidence factors associated with the connections. The time parameters play the role of a first threshold on the time of occurrence of the event; only the events 'older' than the memorised temporal correlation are considered. The unit output is activated when the sum of confidence factors of the different active inputs passes a minimal threshold. 9) The structure of the network is dynamically updated, at each processing cycle. Learning is performed continuously during the experiments, that is we do not distinguish between learning and testing phases. Testing, i.e. retrieving the output occurs at each cycle in order to direct the robot's actions (moving and emission of radio signal).

Figure 1 presents a schematic description of the learning architecture. The learning process used in the experiments is described in the following section.

2.1 Learning Process

Learning is dynamical and performed continuously during the experiment. At each processing cycle, the learner robot's sensor-actuator information is compared to the measurements collected the previous cycle. Whenever a variation in the measurement of one sensor or actuator system is detected by the corresponding *event recognition* module, the new measurement is presented to the associative architecture (DRAMA) to be associated with all simultaneous and previously recorded measurements in the same or other sensor-actuator system. Sensor data are collected as bit-strings (different length for each sensor); in the experiments we considered a one-bit variation in the measurement as an event. Learning occurs only when an *event* has been detected. Events are memorised in the self-recurrent connections of the network input units for a fixed period of time (100 cycles, i.e. about 2.5 minutes of real time on the robot)[1] or until

---

[1] This was preferred to an infinite event memorisation, where e.g. the memory of the event is erased only when a new event is detected, in order to avoid misleading associations. The experimental set-up was done so that the teaching of a word occurred at the latest one or two minutes following the event which the word should be associated with.

a new event occurs. In figure 1, we present a schematic view of the learning architecture. The sensor-actuator systems A, B and C in the figure could be interpreted, for instance, as the motors, compass and radio systems respectively. The learning process works then as follow: If, at time t, one event is detected by the sensory system A and one by the sensory system B, they will be associated with each other. If then, at time $t + n$ ($n <$ memory length), a new event is detected by the system C, it will be associated with the two previously recorded events in systems A and B. Association consists of 1) increase the confidence factors of connections between active units (following an Hebbian rule) by a factor inversely proportional to the temporal delay between activation of each unit 2) update the time parameters to record this temporal delay (mean value over all examples).

2.2 Experimental implementation

Figure 2 shows a schematic description of the control system in the present implementation. In the experiments we report here, the architecture is used for both learner and teacher agent. Basic behaviours of wandering, obstacle avoidance and following are predefined by setting in advance the values of the connection parameters between infra-red sensor, bumpers, light sensors and the agents' motors. Figure 2 shows how obstacle avoidance behaviour with the frontal infra-red sensor was predefined: connections from the frontal infra-red sensor to the motor left are set with predefined values, resulting in the robot turning to the right (activation of the left motor only) when detecting an obstacle by infra-red reflection. Mutual following of the two agents is defined as a phototaxis behaviour performed by both agents with light detectors ([2]) (each agent carrying the relevant sensors on its front (learner) or on its back (teacher)), i.e. asymmetric connections are set between these left-right sensors and right-left motors in order to result in phototaxis behaviour. The robot's behaviour is then determined by retrieving the motor outputs at each cycle given the current sensor input. The motor activity is determined by the activity of the units corresponding to the motor system. The teacher robot's knowledge of the vocabulary is predefined by setting the value of the connections between relevant sensor information and the radio emitter (actuator). Similarly to what is done for the motor activation, the teacher 'speaking' (emission of radio signals) is directed by retrieving the radio output given the current sensor-actuator state. The teacher 'speaks' only when it sees the learner. Finally, learning occurs dynamically as a consequence of the continuous update of the two connection parameters to account for the variations in the sensor-actuator states. Figure 2 shows an example of connection update after training. Training of the network results in associating different compass measurements with different radio signals: 4 signals (external teaching) are associated with measurement of the 4 quadrants (North, South, East, West). Self-organisation of the network results also in associating different compass measurements with different light measurements (we observed in the experiments that when facing the north and so the windows, the robot measures an increase in the global lighting).

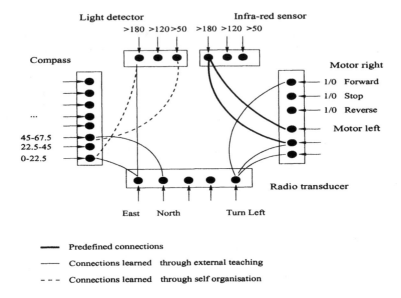

Fig. 2. Schematic representation of the processing of the sensor-actuator information through the DRAMA architecture as implemented in the experiments. The motor activity is determined by retrieving the activity of the units corresponding to the motor system. The motor activity is encoded in a 3-bit string. Bit 1 determines the state of activity of the motor (*active/not* = 1/0), bit 2 encodes for the direction of activity (*forward/reverse* = 1/0) while the third bit determines the speed of activity (*full/half* = 1/0). Basic obstacle avoidance behaviour is predetermined by setting the connection (straight lines) between a high infra-red value detected in infra-red detector I1 and the left motor. As a result, the robot turns to the right in front of an obstacle by activating its left motor forward and keeping the right one inactive. Training of the network results in associating different compass measurements (East, North) with different radio signals (external teaching provided by the teacher agent) and with different light measurements (when facing the north and so the windows, the robot measures an increase in the global lighting).

To summarise, at each processing cycle, given the current sensor input, the actuators' (motors, radio emitter) outputs are calculated. Then, the current sensor and actuator information is compared to the information measured in the previous cycle (in the event detectors). If a variation is noticed, the new information is passed on to the associative memory where it is memorised for a fixed duration, during which it is associated to any incoming event. It is important to understand that the same connections, i.e. a single network, are used for learning, control (motor activity) and signal transmission (radio transmission). In addition, the same learning is used for generating the symbolic associations (radio-sensor) and other sensor-sensor and sensor-actuator associations, and the same retrieving algorithm then use these associations for both controlling the robot's movements and communicative behaviour. Figure 3 shows the variation of activity of the units corresponding to the left and right motors, the com-

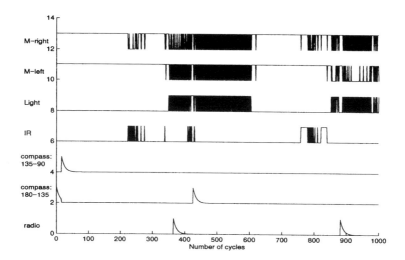

Fig. 3. Variation of activity of the units corresponding to the left and right motor, the compass, light and radio sensors, during 1000 processing cycles. We observe that activation of the infra-red detector unit (at times 210, 350, 430 and 780) produces an immediate deactivation of the right motor. The robot turns to the left when it faces an obstacle, as previously defined by the connection weights. Consequently, after a long rotation a new value is measured by the compass (at time 450). Light detection at time 380, i.e. detecting the other robot, produces a deactivation of left and right motors alternately (time 380-600). The robot aligns behind the other robot.

pass, radio and light sensors, during 1000 processing cycles[2]. We observe that activation of the infra-red detector unit (at times 210, 350, 430 and 780) produces an immediate deactivation of the right motor. The robot turns to the left when it faces an obstacle, as previously defined by the connection weights. Consequently, after a long rotation a new value is measured by the compass (at time 450). Light detection, corresponding to detection of the second robot, produces a deactivation of left and right motors alternately (time 380-600). The robot aligns behind the other robot. We also see the effect of the decrease of activation along the recurrent connections of the radio and compass units. The motor and infra-red units do not decrease because they are constantly activated by the new input. When the radio unit is activated, at cycles 380 and 890, it is associated with the following activation of the compass unit, just before its extinction.

3   The Experiments

In this section, we report on experiments carried out both in simulated and physical environments. Teachings concern the robot's movements (motor states) and

---

[2] Each sensor information is in fact given by more than one unit; in figure 3 we represent the maximal activation of the units corresponding to this system

orientation (compass measurements). Figure 4 presents a schema of the teaching scenario. The learner robot follows the teacher robot that wanders randomly in the arena. Whenever the teacher detects the learner behind it, it stops and sends a radio signal composed of the 'word' *stop* and a word for the *orientation* (*South, North, East, West*) given by its compass. Then it starts to move again and sends consecutively (after a delay of 200 cycles) a teaching signal containing the word *move*. The vocabulary consists at that point of 6 words: [*stop, move, South, North, East, West*]. In a second phase of the experiment, instead of teaching the compass direction three radio signals are sent to distinguish between three movements *stop, turn left* and *turn right*. Each signal is sent separately whenever the teacher performs the particular movement, as part of its wandering, while detecting the presence of the learner behind it. At the end of the experiment, the total vocabulary learned by the learner agent consists then of 8 words: [*stop, move, turn left, turn right, South, North, East, West*].

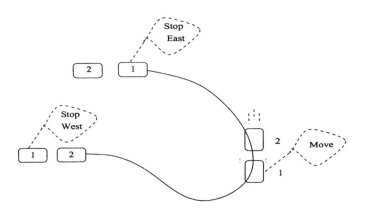

Fig. 4. **The Teaching Scenario.** A teacher robot (1) teaches a learner robot (2) a vocabulary to describe its movements and location. The learner, while following the teacher, associates the teacher's words with its own observations.

Learning consists of multiple associations between all the sensor and motor systems of the robot. More specifically, learning of the language amounts to associate different radio signals (bit-strings) with distinct sensory measurements and motor states. Association are made between all the sensors (light, infra-red, bumpers, compass and radio) and motor systems of the robot, that is associations are not only made between sensor and motor systems but also between the sensor systems themselves, e.g. the radio/compass and radio/light associations. Learning is achieved once the confidence factors associated with the relevant sensory measurements show a strong correlation, easily distinguishable from the ones with other non relevant sensory measurements.

Two autonomous LEGO robots (teacher and learner) are used for the exper-

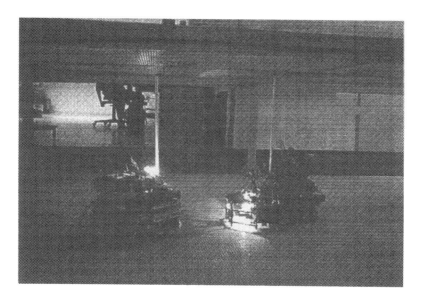

Fig. 5. Left Teacher (in front) and learner robots in their Dodgem environment.

iments. Each robot is equipped with one frontal infra-red sensor and bumpers to avoid the obstacles (see figure 5). They also have two light detectors in the front (learner) or in the back (teacher) to enable them to follow each other. The range and sensitivity of the sensors are given in table 1. They have a radio transceiver, the means of transmission of communication signals. Each signal is encoded in one byte with only 1-bit activated (e.g. 'North' = (10000000), 'Stop' = (00000001)). That is, in our agents' 'language' 8-bit = 1 byte 'words'. The

Sensor type	Sensitivity range	
	Physical Exp.	**Simulations**
Bumpers	0cm (touch contact)	None
Infra-red	cone of 15^0 and 30-40cm	cone of 180^0 and 45cm
Light	cone of 180^0 and 40-50cm	cone of 180^0 and 45cm
Compass	8 quadrants	8 bit-string (22.5^o)
Radio	30m	whole arena

Table 1. Table with the robots sensor sensitivities in physical and simulated environment.

arena consists of a rectangular cage of 2.5m by 2m by 0.5m, in which the robots are continuously recharged similarly to the system used in the 'Dodgem' (bumping cars) game. Roof and bottom of the arena are electrified, creating a potential difference of 10V between them. The robots carry a long stick touching both ends of the cage from which they receive the current to power their battery and light bulb. We estimate that about 10 to 20% of the sensor measurements are noisy: 80% of the radio transmissions are correctly received (i.e. if a signal is received, then it is perfect; the noise corresponds to the case where an emitted signal has not been received), while the quadrants given by the compass are correctly detected in about 90% of the cases ('zigzagging', side movements, of the robots while following each other, and undesirable magnetic effects produced by the motors and the powering of the cage).

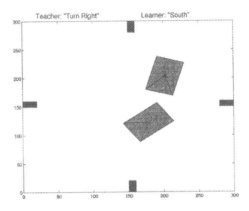

Fig. 6. Simulation studies are carried out in a 2-D simulator, in which two robots can interact in a rectangular arena. In the figure, we see the teacher robot (No 1) followed by the learner robot (2), while the teacher begins to turn to the right and sends the corresponding radio signal. The learner which follows it is still thinking 'move' as it has not begun to turn. The boxes on the side of the arena stand for the obstacles of the real arena.

Simulation studies are carried out in a 2-D simulator, in which two robots can interact in a given environment, here a rectangular arena measuring 300 by 300 units (see figure 6). There are 4 obstacles on the sides of the arena, represented by dark rectangles. The robots are represented as rectangles of 30 by 20 units, a triangle determining the front. The simulated robots are provided with the same sensor systems as the real robots (apart from not having bumpers): light detectors for mutual recognition, infra red sensor for obstacles avoidance, a radio transducer as the means of communication and a compass to measure bearings to the request of 22.5 degrees. Infra-red and light detectors are associated with a cone of vision of 180 degrees, which is segmented into eight quadrants. The measurement of the sensor is given by an 8-bit string where each bit corresponds to the value measured in each of the eight quadrants (e.g. infra-red=(11000000) stands for an infra-red activation of the first two quadrants). The range of sensitivities of the sensors is given in table 1.

The robots' behaviour is calculated separately by the same routines as used in the physical robots. Code is written in C and is processed serially. At each cycle, each robot's routine measures the current sensor-actuator state of the robot. The robots' sensor perceptions are simulated by defining a specific measurement routine for each sensor. In order to reproduce reality more faithfully, the following behaviour is made imperfect. Following is mutual, each agent aligning towards the other one on the basis of its light detectors' measurement (each robot carries a bright light). Similarly to what happens in reality, an agent is able to determine the position of the other agent with respect to its own with a precision of 90 degrees. Therefore, the alignment of the two robots is imprecise. In addition, randomness is introduced in the calculation of the robot's movements, such as to represent the experimental imprecision.

We use a simulator in order to reproduce the physical experiments and study their successes and failures in a more reliable environment. The main advantage of simulations over physical experiments is that they are repeatable, faster (simulating a 1 hour experiment takes about 5 minutes) and do not suffer unexpected hardware breakdown. The disadvantages in terms of model faithfulness are, of course, well known. For a more complete discussion of this see [13]. Here simulation studies are carried out to show the stability and success of the learning in the particular set-up proposed, which is further demonstrated in the physical experiments.

4 Results

A set of 100 runs, each run simulating 10000 processing cycles of the robots (about 4 hours of real experiments) were carried out to test the speed and stability of the learning. The success or failure of learning in the experiment is determined by looking at the values of the connections' confidence factors between the radio sensor system and the rest of the sensor and actuator systems of the agent (light, infra-red, compass, motors). The experiment is said to be successful when only the connections between the correct radio unit, standing for the particular word and its associated sensor (compass) - actuator (motors) unit combination have a confidence factor value greater than the maximal one for all connections leading to these units. In other words, learning occurs if the number of correct associations (correct matching between radio signal reception and sensor-actuator measurement) is greater than the number of incorrect ones, i.e. if the noise (incorrect examples) is below 50% of the total number of examples in a run.

In a Willshaw type of associative memory as used here, there is no notion of convergence as in many other kinds of neural network, since weight update is only of one time step. The success of the learning can be measured at each time step as the ratio between connection parameters. At each presentation of a new example (new teaching), the connections' confidence factors are updated. The study of the variation of the confidence factors values during the experiment informs us about the variation of the percentage of noise and consequently of the stability of the learning. Noise, in our case an incorrect example, is produced when teacher and learner are not perfectly aligned and therefore measure different compass orientations or perform different movements. Figure 7 left shows the mean variation of the confidence factor (cf) values along a run. The y axis represents the ratio between the cf value associated with the correct word and the maximal value of cf attached to all words for a given sensor measurement, i.e. cf(correct correlations)/(cf(correct correlations)+max(cf(incorrect correlations)). The data are the mean value over all words and over all runs. A ratio greater than 0.5 means that the particular sensor measurement has been associated with the correct word more often than with other words. Figure 7 left shows that learning is stable and steadily increasing while more teachings are given. Once the cf value has crossed over the threshold of 0.5, i.e. the word has been learned, it stays over it with a standard deviation of 0.2 (see table 2).

	Simulations Mean & Std	Physic. exp Mean & Std
1	0.62 ± 0.23	0.71 ± 0.18
2	0.80 ± 0.19	0.96 ± 0.10
3	21 ± 31	70 ± 16

Table 2. Comparison between results of simulations and physical experiments. From top to bottom: 1) Ratio between the confidence factor value of the correct word and the maximal value of confidence factor attached to all words for a given sensor measurement (mean value over all words and all runs). 2) Ratio between the number of words learned at the end of the run and the total vocabulary (mean value over all runs). 3) Number of teachings per run (mean value over the runs).

The slow increase of the slope (figure 7 left) at its beginning is due to the fact that we superpose the learning curves of 100 different runs and these do not begin at the same point. Each run starts with the agents in a new random position. The time before the first meeting and teaching varies then for each run. The slope of each single curve is however very steep. In most cases, the first teaching would be a correct example. The curve would then start from the maximal value 1, decreasing slightly afterwards under the effect of noise. This is illustrated by the very steep beginning of the curve after the first teaching.

A set of five physical experiments is reported here. Each physical experiment consisted of two teaching phases, in the first phase six words were taught (*stop, move, South, North, East, West*) and in the second 3 words (*stop, turn left, turn right*). Each experiment lasted for about 1 hour 30 minutes (45 minutes for each phase) and was stopped when about 90 teachings had been done (the exact number of teaching events is imprecise because the radio transmission was imperfect and it was difficult to detect whether the radio signal had been received by the robot.). Table 2 compares results of simulations and physical experiments for learning of the first phase. Results are qualitatively similar. The mean values of confidence factor for all words at the end of each run are respectively 0.62 and 0.71. The standard deviation values represent the fluctuation of the cf values along the run. The qualitative similarity between physical and simulated results shows that the simulations studies gave a good account of the percentage of noise occurring in the physical experiments. Lines 2 and 3 of the table show respectively the ratio to total vocabulary of words that have been learned at the end of the run and the total number of teachings in a run (average values for all runs)[3]. Surprisingly, the results of the physical experiments are slightly better than the simulations. A reason may be the small numbers of the physical experiments. But the main reason is surely that learning in the physical experiments was helped by placing the robots one behind the other one from the beginning[4]. The chances of the two robots meeting and then teaching were therefore greatly increased. This may explain why the number of teachings and the number of learned words per run is higher in the physical experiments than in the simulations.

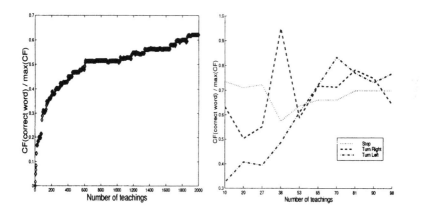

Fig. 7. Variation of confidence factor values along a run in simulation studies (left) and physical experiments (right). Y axis represents the ratio between the cf value of the correct word and the maximal value of cf attached to all words for a given sensor measurement. Data are mean value over all words and over all runs. A ratio greater than 0.5 means that the particular sensor measurement has been associated with the correct word more often than with other words.

In figure 7 right, we show the variation of the confidence factor values in the second phase of the experiments. We observe that the three words are correctly learned after 30 teachings, which corresponds to about 15 to 20 minutes of

---

[3] The large value of standard deviation in line 3 is due to outlier values of 300 (teachings per run) in the data.

[4] This was done in order to shorten the time of the experiments, the hardware being unlikely to stand more than one hour of continuous running of the robots.

experiment, which is faster than in the simulations (figure 7 left). Again, this is probably due to our 'helping' the physical robots to align faster. However, the fluctuation of the values are more pronounced in the physical experiments, which means that the proportion of noise varies more in the physical world. The noise corresponds to the cases where the radio signals are associated with incorrect motor states or compass measurements. Because the movements of the physical robots are less smooth, incorrect alignment and imprecise following of the two robots is more pronounced and more frequent, which results in the robots facing and measuring different directions and producing different movements (e.g. turning in opposite directions). Note also that learning of the word *stop* is better and occurs already from the beginning, due to the fact that this word had already been taught during the first phase of the experiment (no erase of the memory between the two phases).

In summary, the experiments showed that 1) a number of words are learned and correctly associated; 2) the vocabulary is learned although the experimental conditions create incorrect teachings, which demonstrates the robustness of the learning architecture in the face of 20 percent of noise; 3) the learning is stable once a significant number of relevant correlations have been recorded. One may now question whether these experiments, because of their apparent simplicity (the vocabulary has only eight words), are a demonstration of the efficacy of the learning method. Although the sensory system and consequently the vocabulary are quite simple, the learning task is actually rather difficult. Because of their very limited sensory capabilities, the robots' mutual recognition and following is very imprecise. Alignment of the two robots is seldom perfect. Consequently, the chances of their sensor measurements being identical (in particular their compass measurements) are quite slim and misleading associations may easily occur. On the contrary, if we had used more sophisticated sensors, giving a finer and thus more informative description of the environment, the number of taught words could have been greatly increased. On the other hand, however, we would have simplified the task of imitation by giving the robots more information on which to rely for adjusting their positions one towards the other. This leads us to think that the levels of difficulty of the learning task in the two cases are in some sense comparable. The significant data have to be extracted from a considerable amount of noise that is produced, in the first case, by misleading teachings due to an imprecise imitation and, in the second case, by a more rich source of information. Therefore, it seems reasonable to conclude that our experiments, although apparently simple, are relevant as a demonstration of the validity of the system.

5 Conclusion

This paper reported on experiments where a physical robot is taught a vocabulary concerning proprio and exteroreceptive perceptions. Teaching is provided by a second heterogeneous robot using an unsupervised imitative teaching strategy, namely mutual following of the two agents. Grounding of the vocabulary occurs as part of a general associative process of all the robot's sensor-actuator measurements. Simulations and physical experiments are carried out which show successful and consistent results. The vocabulary is learned and learning is stable in the face of a significant amount of noise. Success of the experiments demonstrates the validity of the learning architecture for the particular task. This work suggests that a simple movement imitation strategy is an interesting scenario for the transmission of a language, as it is an easy means of getting the agents to share the same context of perceptions, a prerequisite for a common understanding of the language to develop. This work may be a starting point towards using more complex imitative strategies for transmitting more complex forms of communication.

6   Acknowledgement

An enormous thanks to the technicians for their essential support. Many thanks to John Demiris and Auke Jan Ijspeert for their useful comments on this paper. Aude Billard is supported by a personal grant from the LEMO company and the Swiss National Research Fund.

References

1. T. Arita, C.E. Taylor: A Simple Model for the Evolution of Communication. The Fifth Annual Conference on Evolutionary Programming (1996).
2. A. Billard, G. Hayes: A Dynamical Recurrent Associative Memory Architecture for Learning Temporal Correlated Events in Autonomous Mobile Agents. To appear in *Adaptive Behaviour* journal, MIT Press (1998).
3. A. Billard, K. Dautenhahn: Grounding communication in autonomous robots: an experimental study. To appear in *Robotics and Autonomous Systems*, special issue on Quantitative Assessment of Mobile Robot Behaviour. Edited by U. Nehmzow, M. Recce, and D. Bisset (1998).
4. J. Demiris, G. Hayes: Imitative Learning Mechanisms in Robots and Humans. Proc. of the 5th European Workshop on Learning Robots, Bari, Italy (1994).
5. K. Dautenhahn: Getting to know each other: artificial social intelligence for autonomous robots. Robotics and Autonomous Systems, vol. 16, pages **333-356** (1995).
6. V. Klingspor, J. Demiris and M. Kaiser: Human-Robot-Communication and Machine Learning. Applied Artificial Intelligence Journal, Vol. 11, to appear (1997).
7. Y. Kuniyoshi, M. Inaba and H. Inoue: Learning by Watching: Extracting Reusable Task Knowledge from Visual Observation of Human Performance. In IEEE Transactions on Robotics and Automation', Vol. 10, No 6, Dec. 1994.
8. Mataric M. J.: Reducing Locality Through Communication in Distributed Multi-Agent Learning. To appear in the *Journal of Experimental and Theoretical Artificial Intelligence*, special issue on Learning in Distributed Artificial Intelligence Systems, Gerhard Weiss, ed., fall 1997.
9. B. MacLennan & Gordon M. Burghardt: Synthetic Ethology and the Evolution of Cooperative Communication, Adaptive Behavior, Vol. 2, No. 2, pp. 161-187, Fall 1993.

10. M. Oliphant: Formal Approaches to Innate and Learned Communication: Laying the Foundation for Language. Doctoral dissertation, Department of Cognitive Science, University of California, San Diego (1997).

11. L. Steels & P. Vogt: Grounding adaptive language games in robotic agents, Proceedings of the fourth European conference on Artificial Life, Brighton, U.K., July 1997.

12. L. Steels: The Spontaneous Self-organization of an Adaptive Language. In Muggleton, S. (ed.) *Machine Intelligence* 15. Oxford University Press, Oxford (1996).

13. M. C. Torrance: The Case for a Realistic Mobile Robot Simulator. In Working Notes of the AAAI Fall Symposium on Applications of Artificial Intelligence to Real-World Autonomous Mobile Robots, Cambridge, MA, October 1992.

14. H. Yanco & A. S. Lynn: An adaptive communication protocol for cooperating mobile robots. In From Animals to Animats 2: Proceedings of the Second International Conference on the Simulation of Adaptive Behavior, edited by J.-A. Meyer, H.L. Roitblat and S.W. Wilson. The MIT Press/Bradford Books, pp. 478-485 (1993).

Continual Robot Learning with Constructive Neural Networks

Axel Großmann and Riccardo Poli

School of Computer Science, The University of Birmingham, UK
{A.Grossmann, R.Poli}@cs.bham.ac.uk

Abstract. In this paper, we present an approach for combining reinforcement learning, learning by imitation, and incremental hierarchical development. We apply this approach to a realistic simulated mobile robot that learns to perform a navigation task by imitating the movements of a teacher and then continues to learn by receiving reinforcement. The behaviours of the robot are represented as sensation-action rules in a constructive high-order neural network. Preliminary experiments are reported which show that incremental, hierarchical development, bootstrapped by imitative learning, allows the robot to adapt to changes in its environment during its entire lifetime very efficiently, even if only delayed reinforcements are given.

1 Introduction

The development of learning techniques for autonomous robots constitutes one of the major trends in the current research on robotics [2]. Adding learning abilities to robots offers a number of benefits. For example, learning is essential when robots have to cope with dynamic environments. Moreover, it can help reduce the cost of programming robots for specific tasks. These and other features of robot learning are hoped to move autonomous robotics closer to real-world applications.

Reinforcement learning has been used by a number of researchers as a computational tool for constructing robots that improve themselves with experience [6]. Despite the impressive advances in this field in recent years, a number of technological gaps remain. For example, it has been found that traditional reinforcement learning techniques do not scale up well to larger problems and that, therefore, 'we must give up *tabula rasa* learning techniques' [6, p. 268] and guide the learning process by shaping, local reinforcement signals, imitation, problem decomposition, and reflexes. These techniques incorporate a search bias into the learning process which may lead to speed-ups. However, most of the proposed solutions do not address sufficiently the issue of continual adaptation and development.

If a robot is provided with a method for measuring performance, learning does not need to stop. Robots could adapt to changes in their environment during their entire lifetime. For example, Dorigo and Colombetti [4] have proposed an incremental development approach in which a learning agent goes through three stages of development during its life: a 'baby phase', a 'young phase', and an 'adult phase'. In the 'adult phase', a monitoring routine is used, which can reactivate either the trainer (used in the 'baby phase') or the delayed-reinforcement modules (used in the 'young phase') if the agent's performance drops.

Our approach to continual learning is slightly different. We believe that incremental, hierarchical development is essential for continual learning. If a robot has to learn a new behaviour during its lifetime, it should make use of the previously learned behaviours. Machine learning mechanism that could allow this include the default hierarchies produced in learning classifier systems [11] and the automatic addition of units and connections in constructive neural networks [10]. Of particular interest in the field of constructive networks are the Temporal Transition Hierarchies introduced by Ring [10] who has shown that they can be used as supervised learning algorithms in a Q-learning reinforcement system.

The aim of this paper is to show that temporal transition hierarchies can be used for the integration of several robot-learning techniques. We propose an approach for combining reinforcement learning, imitation, and incremental hierarchical development. We apply this approach to a simulated mobile robot that learns to perform a navigation task by imitating the movements of a teacher and then continues to learn by receiving reinforcement.

The paper is organised as follows. In Section 2, we describe the learning scenario which we have chosen for our experiments. In Section 3, we introduce the particular connectionist model we are using for learning the robot skills. In Section 4, we explain how the robot can acquire skills hierarchically and incrementally by imitative learning. In Section 5, we explain how skills learned by imitation can be used in a reinforcement-learning system. In Section 6, we demonstrate how additional behaviours can be learned by receiving delayed reinforcement. Finally, we draw some conclusions in Section 7. We explain our ideas using examples obtained from our actual experiments; there is no particular section for the discussion of experimental results. Future papers we be devoted to a fuller validation of our approach.

2 The Learning Task

For a robot to learn a task, it must be given some feedback concerning its performance. In reinforcement learning, the robot is given a scalar performance signal (called reinforcement) telling it how well it is currently doing at the task. The success of the learning process depends very much on how often the robot receives rewards. It may receive reinforcement after each action or after a sequence of actions. For many learning tasks, only delayed reinforcement is available. For example, imagine a robot wandering from a start state to a goal state. It will receive no positive feedback at all during most of its interaction with the environment. Reinforcement will only be given when it actually reaches the goal. This example illustrates why learning using delayed reinforcement is usually difficult to achieve. If the task is complex and the robot has to learn from scratch, then the robot is unlikely to find any action for which it is given reward.

For our experiments, we have chosen a learning task where only delayed reinforcement is available. A mobile robot has to travel through a maze-like environment. The robot always starts from a predefined position. By reacting correctly to the corners, it has to find its way to a specified goal position. To receive reinforcement, the robot has to reach the goal area and to stop there.

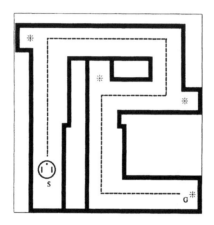

(a) Maze-like environment.

$$no\_light(t) \wedge light\_left(t-1) \rightarrow turn\_right$$
$$no\_light(t) \wedge light\_right(t-1) \rightarrow turn\_left$$
$$no\_light(t) \rightarrow move\_forward$$
$$light\_left(t) \rightarrow turn\_right$$
$$light\_right(t) \rightarrow turn\_left$$
$$light\_ahead(t) \rightarrow stop$$

(b) Default hierarchy for maze navigation.

Fig. 1. Learning a navigation task with delayed reinforcement.

The experiments have been performed using a physically realistic simulator for the Khepera robot [9]. Khepera is a miniature mobile robot equipped with eight infrared sensors, six at the front and two at the back. Using its sensors, the robot can detect obstacles and light sources within a range of 40 mm. A typical test environment is shown in Fig. 1(a) where the start and goal positions are marked with S and G, respectively. In some corners of the maze, there are light sources which can be detected by the robot.

Learning from delayed reinforcement is generally a slow process. Therefore, a straightforward application of reinforcement-learning techniques is often not practical. To speed up the learning process, we have used task decomposition and abstraction. Namely, we have predefined a small set of medium-level motor and pattern-recognition behaviours which enormously reduce the size of the search space the learning algorithm has to explore. We have restricted the number of actions the robot can perform by defining four motor controllers: *move_forward*, *turn_left*, *turn_right*, and *stop*. That is, the robot can follow a path, perform 90-degree turns in front of obstacles, and stop at any position. Moreover, the robot can distinguish the following light-configurations: *light_ahead*, *light_right*, *light_left*, and *no_light*.

The motor controllers have been implemented as hand-written routines whereas the light-detection skills have been learned off-line using a feed-forward neural network and a supervised learning algorithm. Sensor readings for typical light-configurations have been used as training instances.

In the given learning task, the robot has four different sensations and four actions. It has to learn to choose the right action each time there is a wall blocking its way. That is, the robot has to solve a sequential decision task. This could be done by developing a set of sensation-action rules. Preferably, the rule set should be minimal. A way of achieving this is to design a default hierarchy in which general rules cover default situations, and specialised rules deal with exceptions like the rule set in Fig. 1(b). In the next sections, we will show how such rule sets can be represented and learned by a constructive neural network.

3 Transition Hierarchy Networks

Ring [10] has developed a constructive, high-order neural network, known as Temporal Transition Hierarchies, that can learn continuously and incrementally. There are two types of units in the network: *primitive* units and *high-level* units. The primitive units are the input and output neurons of the network and represent the sensations and actions of the learning agent. The high-level units enable the network to change its behaviour dynamically. More precisely, a high-level unit l_{ij} is assigned to the synaptic connection between the lower-level units i and j and can modify the strength w_{ij} of that connection.

When no high-level units are present, the transition hierarchy network behaves like a simple feed-forward neural network. The activation of the i-th action unit is the sum of the sensory inputs, multiplied by the high-order weight $\hat{w}_{ij}$ connected to unit i. The activation of the high-order units, if present, is computed in the same way. The high-order weights $\hat{w}_{ij}$ are defined as follows. If no l unit exists for the connection from j to i, then $\hat{w}_{ij} = w_{ij}$. If there is such a unit, its previous activation value is added to w_{ij}, i.e., $\hat{w}_{ij}(t) = w_{ij} + l_{ij}(t-1)$.

The function of the high-level units will be illustrated with an example. The transition hierarchy network shown in Fig. 2 solves the navigation task in

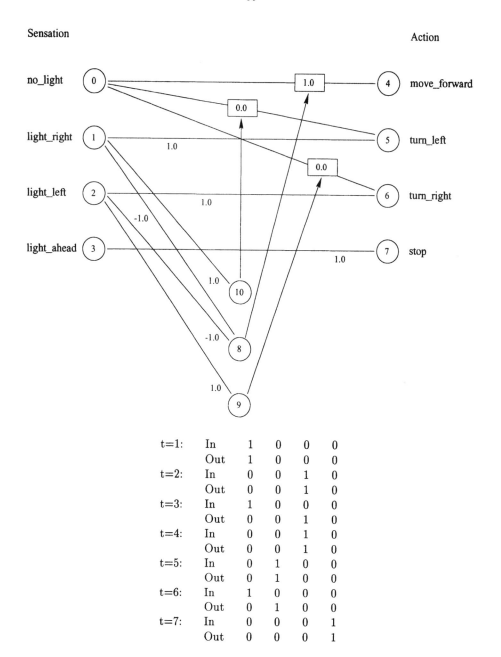

Fig. 2. Transition hierarchy network to solve the navigation task (above) and the outputs produced at different time steps (below).

Fig. 1. As explained in Sect. 4, it has been created automatically. Until it reaches the goal position, the robot has to change its direction of travel several times. The optimal sequence of decisions can be described through sensation-action rules. Indeed, the network is a computational representation for the rule set in Fig. 1(b). The net outputs are computed each time the robot detects an obstacle in its way. Therefore, we distinguish several time steps t. In time step 2, 4, 5, and 7, the robot can select the action on the basis of the current sensation only. In the absence light however, the decision on which action to take depends also on the sensation in the previous time step. This ambiguity (called hidden state problem) is solved using the high-level units 8, 9, and 10. For example, the high-level unit 8 inhibits the default rule $\langle$no_light$(t) \rightarrow$ move_forward$\rangle$ if *light_right* or *light_left* was active in the previous time step.

The connection weights of temporal transition hierarchies can be determined using a supervised learning technique. Ring [10] has derived a learning rule that performs gradient descent in the error space. Transition hierarchy nets learn constructively and incrementally; new high-level units are created while learning. A new unit needs to be added when a connection weight should be different in different circumstances, see [10, p. 59]. Therefore, long-term averages are maintained for every connection weight. Ring's experimental results show that the learning algorithm performs very well compared to other neural-network models for temporal processing.

4   Learning by Imitation

Recently, several researchers have proposed to use imitation as a way for robots to learn new skills [1, 3, 5, 7]. The main idea is that the robot learns how to act by perceiving and imitating the actions of a teacher. In that way, the robot can find out what is a good action to perform under which circumstances.

We decided to study the possibilities offered by imitative learning, in particular, the idea that the knowledge of successful action sequences reduces the size of the search space that the robot would need to explore in order to learn new skills using reinforcement learning. Our learning scenario is similar to that used by Hayes and Demiris [5]. Two robots are present in the environment: a teacher and a learner. The teacher robot is travelling through a maze along the optimal path with the learner robot following at a certain distance. The learner detects and records 'significant events'. That is, it notices when the teacher changes its direction of travel in front of a corner. The learner then associates the current light-configuration with the action the teacher has carried out, and imitates that action.

To learn by imitation, the learner robot needs the following additional skills: (1) the learner has to be able to follow the teacher; (2) the learner has to be able to detect changes in the movement of the teacher, specifically, to detect 90-degree turns; and (3) the learner has to associate an action with the light-configuration currently sensed. One can imagine many possible realisations of the teacher-following behaviour and the turn-detection skill. For the former, we have used a simple pattern-associator neural network which estimates the current distance from the teacher on the basis of the current sensor readings. The latter is performed by a partial recurrent neural network. Both networks have been trained off-line using typical sensor readings.

To detect a significant event, the learner needs, of course, a corresponding recognition behaviour. In our experiments, the segmentation of significant events is performed by the predefined pattern-recognition behaviours (see Sect. 2).

The situation-action association is actually the crucial part in our imitative-learning approach. The selection of an optimal computational mechanism for this task depends on the amount of information to be stored, the structure of the learning tasks, and the duration of the learning process. Given that in our experiments sensations are pre-processed, that the number of possible actions is finite, and that knowledge reuse can help the learning process, then temporal transition hierarchies seemed very well suited to perform this task.

So, the sensation-action association behaviour was implemented as follows. The learner robot is watching the teacher on its way through the maze, recording the actions performed and light-configurations detected. When the tour is finished, i.e., the goal position has been reached, the sensations and actions stored are used to train a transition hierarchy network. For each significant event, there is a sensation-action pair, which is used as a training instance by the supervised learning algorithm. If a light-configuration has been present, then the activation of the corresponding sensory unit is set to 1.0, to 0.0 otherwise. The activation values of all sensory units form the input vector for the network. The target vector contains an activation 1.0 for the action chosen; all other action units are required to be inactive, i.e., their activation is 0.0. The input and target vectors are presented several times to the network. The sequence of training instances is fixed and corresponds to the temporal order of the significant events.

In our experiments, the learning algorithm did converge very quickly (in less than 30 training epochs). The transition hierarchy network shown in Fig. 2 has been obtained using imitative learning (the connection weights have been rounded to the closed integer after learning). Theoretically, it is possible to collect more than one set of associations and learn them altogether using a single network.

5 Bootstrapping Reinforcement Learning

The reproduction of the teacher's actions in itself cannot solve the problem of continual learning as soon as the teacher robot is removed. Reinforcement learning could but it is computationally impractical. We believe that one can overcome the practical problems of traditional reinforcement learning techniques by gaining information about successful action sequences through imitating an expert on one hand, and by reusing previously learned sensation-action rules on the other hand.

To explore these ideas, we have decided to use Q-learning which is probably the most popular and well understood model-free reinforcement-learning algorithm [6]. The idea of Q-learning is to construct an evaluation function $Q(s, a)$, called Q-function, which returns an estimate of the discounted cumulative reinforcement, i.e., the utility, for each state-action pair (s, a) given that the learning agent is in state s and executes action a. Given an optimal Q-function and a state s, the optimal action is found simply by choosing the action a for which $Q(s, a)$ is maximal. The utility of doing an action a_t in a state s_t at time t is defined as the expected value of the sum of the immediate reinforcement r plus a fraction γ of the utility of the state s_{t+1}, i.e., $Q(s_t, a_t) = r(s_t, a_t) + \gamma(\max_a Q(s_{t+1}, a))$, with $\gamma \in [0, 1]$.

A transition hierarchy network can be used to approximate the utility function $Q(s, a)$. The states s can be represented by the activation pattern of the sensory units, whereas the utility $Q(s, a)$ can be represented by the activation value of the corresponding action unit in the network. After learning by imitation, the outputs of the network represent actions and not utilities. Nevertheless, we can use the outputs as initial utilities, i.e., as starting points for the computation of the correct utility values. To use the net output as utilities, the output of the action units has to be discounted according the action's contribution to achieving reinforcement. One way to perform this transformation is to let the robot receive reinforcement from the environment and to apply Lin's algorithm for connectionist Q-learning [8]. The algorithm uses the following rule to update the stored utility values: $\Delta Q(s_t, a_t) = r(s_t, a_t) + \gamma(\max_a Q(s_{t+1}, a) - Q(s_t, a_t))$ where $\Delta Q(s_t, a_t)$ is the error value corresponding to the action just performed.

Generally, a transition hierarchy network will grow while learning the Q-values since a single sensation-action rule can have different utilities values at different time steps if utilities are discounted, i.e., if $\gamma < 1$. The effect of Lin's algorithm on the network in Fig. 2 for $\gamma = 0.91$ can be seen in Fig. 3. Two high-level units have been added during the learning process to distinguish the utility values of the rule $\langle$light_left$(t) \rightarrow$ turn_right$\rangle$ in time step 2 and 4. Since Lin's algorithm changes the utility values only for actions which have actually been performed, the output of some action units becomes negative. However, these values could be changed into 0.0 if necessary. Provided that the environment has not changed, the Q-learning algorithm converges very quickly in less than 15 trials, a trial being a repetition of the navigation task followed by receiving reinforcement each time at the goal position.

6 Learning Additional Behaviours

After a robot has learned an optimal behaviour, the environment might change. If the robot is not able to adapt to the new situation, it will keep trying formerly effective actions without any success. The objective of our continual-learning approach is to allow the robot to adapt to environmental changes while using as much previously learned knowledge as possible to find a solution for the current task.

The sensation-action rules represented in the transition hierarchy network need revision after changes in the environment or in the learning task. Some

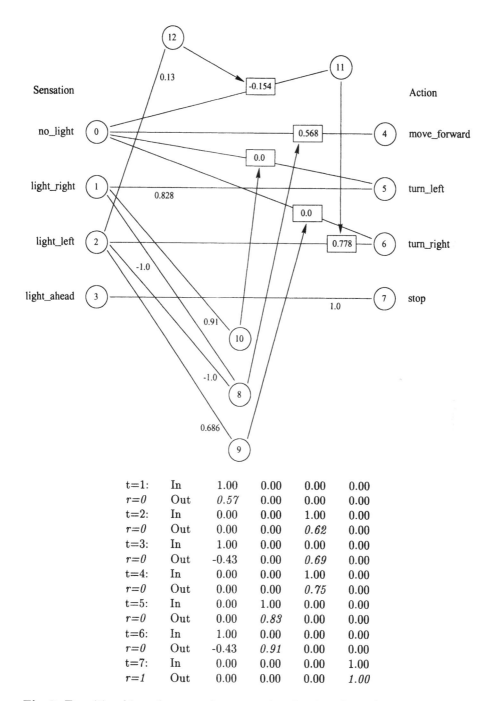

t=1:	In	1.00	0.00	0.00	0.00
r=0	Out	0.57	0.00	0.00	0.00
t=2:	In	0.00	0.00	1.00	0.00
r=0	Out	0.00	0.00	0.62	0.00
t=3:	In	1.00	0.00	0.00	0.00
r=0	Out	-0.43	0.00	0.69	0.00
t=4:	In	0.00	0.00	1.00	0.00
r=0	Out	0.00	0.00	0.75	0.00
t=5:	In	0.00	1.00	0.00	0.00
r=0	Out	0.00	0.83	0.00	0.00
t=6:	In	1.00	0.00	0.00	0.00
r=0	Out	-0.43	0.91	0.00	0.00
t=7:	In	0.00	0.00	0.00	1.00
r=1	Out	0.00	0.00	0.00	1.00

Fig. 3. Transition hierarchy network representing Q-values (above) and the outputs produced at different time steps (below).

of the rules might need to be retracted and possibly replaced by new rules. The revision should fulfil the following three criteria: (1) the rule set should be minimal; (2) the amount of information lost in the revision should be minimal; and (3) the least useful rules should be retracted first, if necessary. These criteria could be used to evaluate the performance of any continual learning algorithm.

Ring has performed an experiment with temporal transition hierarchies to investigate their capability of reusing previously learned knowledge [10]. He found that the network was actually capable of taking advantage of and building onto results from earlier training. These features, however, have not been tested in real-world applications. For example, we have found that transition hierarchy networks are very sensitive to noise while learning. Under these circumstances, more units are created than actually necessary.

In our experiments with changing environments, we have adopted a rather pragmatic approach which differs from traditional reinforcement-learning techniques. To keep the size of the network as small as possible, the learning algorithm is made to learn correct utility values only for action sequences which have been found to be useful.

---

1. Create a network TTH_{STM} as a copy of TTH_{LTM}.
2. Set robot to start position; reset the activations of TTH_{STM} and TTH_{LTM}; clear the memory from experiences.
3. Provide the pre-processed sensations as input s to TTH_{STM} and TTH_{LTM}; propagate the activations.
4. Select an action a probabilistically based on the output of either TTH_{STM} or TTH_{LTM}.
5. Perform the action a; get reinforcement r; store the experience (s, a, r).
6. If $(r > 0)$ then replay stored experiences for TTH_{LTM} until its outputs have converged and suspend learning process.
7. Adjust TTH_{STM} by back-propagating $\Delta Q(s, a)$.
8. If $(a = stop)$ go to 2 else go to 3.

Fig. 4. Algorithm for continual reinforcement learning with temporal transition hierarchies.

To keep the changes in the rule set minimal, two networks are used during the search for a successful action sequence. One transition hierarchy network, denoted as TTH_{LTM}, serves as long-term memory and another network, denoted as TTH_{STM}, is used as a short-term memory. TTH_{LTM} learns from positive experience only, whereas TTH_{STM} keeps track of unsuccessful actions produced by the network. The steps performed during the search for a solution are given in Fig. 4. The robot has to find a solution for the changed learning task by trial-and-error interaction with the environment. It starts a trial always at the predefined start position (step 2). The actions are chosen probabilistically (step 4) as follows.

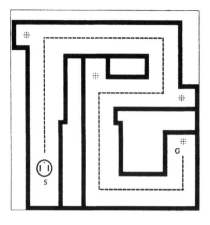

$$\text{no\_light}(t) \wedge \text{no\_light}(t-1) \rightarrow \text{turn\_left}$$

Fig. 5. Changed test environment requiring to learn a new rule for the final turn.

With a certain probability the robot takes the utility values provided by either TTH_{LTM} or TTH_{STM}. There is always a minimal probability for any action to be chosen. The probability of selecting an action is proportional to its utility value. The utility values are updated (steps 6 and 7) according to the rule described in the previous section with the following exception. The utility of the selected action must not be smaller than the utilities of the other actions. Therefore, the utility values of non-selected actions are reduced if necessary. Each time a solution has been found, TTH_{LTM} is trained (step 6) using the successful action sequence which has been recorded (step 5), and TTH_{STM} is replaced with a copy of the new long-term memory network.

We have performed several experiments in which the robot had to adapt its behaviour to changes in the environment. We have created new environments by changing the position of walls, or just by modifying the start or goal position of the robot. After finding a solution, the robot had to revise the utilities stored in the network. A simple example is shown in Fig. 6 where the robot had to learn an additional rule to respond to the changed goal position in Fig 5. By applying Ring's supervised learning algorithm to the TTH_{LTM} network (step 6 in Fig. 4), a new link between the units 0 and 10 was added and the weights were adjusted to represent the new Q-values.

In our experiments, we used a neural network with a fixed number of input and output neurons which correspond respectively to the predefined pattern-recognition and motor behaviours mentioned in Sect. 2. We believe that these basic behaviours can be learned as well, for example, using unsupervised learning techniques. In this way, the predefined behaviours would correspond to the behaviours built up in Piaget's exploration mechanism, namely, a set of 'base behaviours' from which to draw on to produce imitatory behaviours, and a set of 'registrations' with which to compare and recognise behaviours, see [1]. More-

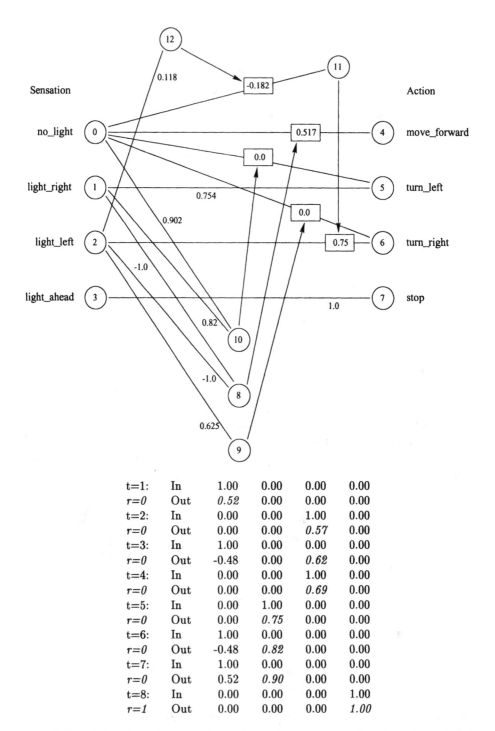

Fig. 6. Transition hierarchy network adapted to the environment in Fig. 5 (above) and the outputs produced at different time steps (below).

over, it should be noted that new input and output neurons can be added during the learning process, if necessary. Adding sensory and action units does not affect previously learned behaviours.

In the experiments so far, we were mainly interested in the changes of the network structure during the learning process. Therefore, a simple learning task has been chosen which can be solved using a rather simple set of rules. The robot learns to find its goal position by solving a sequential decision task. Despite the simplicity of the learning task and the absence of contradictory rules, the convergence of the framework is provided by the constructive character of the learning algorithm. On the one hand, a contradicting rule will be revised, if necessary, in step 6 of the learning algorithm (see Fig. 4). On the other hand, as much as possible of the previously learned behaviours is kept by adding a new unit which makes use of the temporal context, see [10, ch. 6].

7 Conclusions and Future Work

We have proposed an approach in which a mobile robot learns skills hierarchically and incrementally by imitative learning. The skills are represented as sensation-action rules in a constructive high-order neural network. We have demonstrated that incremental, hierarchical development, bootstrapped by imitative learning, allows the robot to adapt to changes in its environment during its entire lifetime very efficiently, even if only delayed reinforcements are given.

We think that the usefulness of incremental learning might depend on the specific learning tasks to be carried out. For example, it might only be useful if the learning tasks are correlated. The learning tasks used in our experiments were simple but typical for autonomous robots. In the future, we will apply our approach to more complex learning tasks on a real robot and we will investigate in which situations incremental learning is beneficial.

Acknowledgements

The authors wish to thank Olivier Michel, who has developed the Khepera Simulator [9] used in the experiments, and the members of the EEBIC (Evolutionary and Emergent Behaviour, Intelligence and Computation) group for useful discussions and comments. Also, thanks to an anonymous reviewer for many useful suggestions in improving this paper.

Bibliography

[1] Paul Bakker and Yasuo Kuniyoshi. Robot see, robot do: An overview of robot imitation. In *Proceedings of the AISB'96 Workshop on Learning in Robots and Animals*, Brighton, UK, 1996.

[2] Jonathan H. Connell and Sridhar Mahadevan, editors. *Robot Learning.* Kluwer Academic, Norwell, MA, USA, 1993.

[3] John Demiris and Gillian Hayes. Imitative learning mechanisms in robots and humans. In Volker Klingspor, editor, *Proceedings of the Fifth European Workshop on Learning Robots*, Bari, Italy, 1996.

[4] Marco Dorigo and Marco Colombetti. The role of the trainer in reinforcement learning. In S. Mahadevan *et al.*, editors, *Proceedings of the Workshop on Robot Learning held as part of the 1994 International Conference on Machine Learning (ML'94) and the 1994 ACM Conference on Computational Learning Theory (COLT'94)*, pages 37–45, 1994.

[5] Gillian Hayes and John Demiris. A robot controller using learning by imitation. In *Proceedings of the Second International Symposium on Intelligent Robotic Systems*, Grenoble, France, 1994.

[6] Leslie Pack Kaelbling, Michael L. Littman, and Andrew W. Moore. Reinforcement learning: A survey. *Artificial Intelligence Research*, 4:237–285, 1996.

[7] Yasuo Kuniyoshi, Masayuki Inaba, and Hirochika Inoue. Learning by watching: Extracting reusable task knowledge from visual observation of human performance. *IEEE Transactions on Robotics and Automation*, 10(6), 1994.

[8] Long-Ji Lin. Self-improving reactive agents based on reinforcement learning, planning and teaching. *Machine Learning*, 8(3/4):293–321, 1992. Special Issue on Reinforcement Learning.

[9] Olivier Michel. *Khepera Simulator Package 2.0.* University of Nice Sophia-Antipolis, Valbonne, France, 1996. Available via the URL http://wwwi3s.unice.fr/~om/khep-sim.html.

[10] Mark Bishop Ring. *Continual learning in reinforcement environments.* PhD thesis, University of Texas, Austin, TX, USA. Available via the URL http://www-set.gmd.de/~ring/Diss/, 1994.

[11] R. L. Riolo. The emergence of default hierarchies in learning classifier systems. In J. D. Schaffer, editor, *Proceedings of the Third International Conference on Genetic Algorithms*, pages 322–327. Morgan Kaufmann, 1989.

Robot Learning and Self-Sufficiency: What the Energy-Level Can Tell Us about a Robot's Performance

Andreas Birk

Vrije Universiteit Brussel, Artificial Intelligence Laboratory,
Pleinlaan 2, 1050 Brussels, Belgium,
cyrano@arti.vub.ac.be, http://arti.vub.ac.be/~cyrano

Abstract. Electrical energy is an important factor for a robotic agent when it tries to stay autonomously operational for some time. Monitoring this variable can provide important feedback for learning. In this paper, we present two different learning criteria based on this idea. Dealing with self-sufficient agents, i.e., agents that roughly speaking have a job to do, one criterion works over cycles of iterated "work" and "recovery". In doing so, it gives some kind of feedback of the robot's efficiency. We argue that a second criterion is needed for learning of most basic behaviors as well as in emergency situations. In these cases, fast and strong feedback, somehow comparable to pain, is necessary. For this purpose, changes in the short-term energy-consumption are monitored. Results are presented were basic behaviors of a robot in a a real-world set-up are learned using a combination of both criteria. Namely, the robot learns a set so-called couplings, i.e., combinations of simple sensor-processes with elementary effector-functions. The couplings learned enable the robot to do touch-based as well as active IR obstacle-avoidance and autonomous recharging on basis of phototaxis.

1   Introduction

Herbert H. Simon defined learning [Sim83] as "any change in a system that allows it to perform better the second time on repetition of the same task or another task drawn from the same population". Marvin Minsky put it similarly as "Learning is making useful changes in our mind"[Min85]. A crucial problem is to find a concrete meaning for "useful" and "good performance".

When following the perspective of Artificial Life on AI [Ste94a] or the animat-approach [Wil91], a key design-aspect of agents is *self-sufficiency*, i.e., "they have to able to sustain themselves over extended periods of time"[Pfe96]. An agent who manages autonomously to stay alive, in the sense of staying operational, performs obviously good, whereas death, in the sense of running out of energy or breaking apart, is bad performance. Ashby formalized this idea as early as 1952 by introducing the notion of *essential variables*[Ash52]. These are the state variables which insure successful operation as long as they are kept within the crucial boundaries or the *viability zone* of the agent's state space.

Due to its "universal" character, the notion of "try to stay alive" can be a general basis for building learning criteria for autonomous agents. At first glance, this idea can be used straightforward. But there is a vicious circle involved: we want the agent to learn to survive (better), but in doing so, we want him not to die. So, any informational feedback seems impossible. The moment the agent gets the crucial feedback, it is already too late, because he is dead. If we do not want to use a mechanism similar to genetic evolution, but "real" learning in the sense of adaption "within" a single agent, we have to look for mechanisms to estimate the agent's capability of keeping viable.

Electrical energy is one of the most important factors for the viability of autonomous robotic agents. In this paper, we investigate how the energy-level can concretely be monitored to judge the agent's well-doing and how this information can be used for learning. As we focus on self-sufficient agents, i.e., agents with a "job to do" to "earn their living", we discuss one criterion working over cycles of iterated "working" and "recovery" following David McFarlands and Emmet Spiers notion of basic cycles [MS97]. This criterion indicates when an agent works more efficient. It works well in simulations where an autonomous mobile robot learns basic behaviors like obstacle-avoidance, phototaxis towards a "foodsource", and so on. But it is not sufficient for real-world experiments. As we will see, a pain-like criterion is needed in addition giving fast and significant feedback in critical situations.

The rest of the paper is organized as follows. Section 2 describes a set-up where robotic agents can "generate" and access "food" in form of electrical energy. We will argue that this so-called ecosystem provides general insight as every self-sufficient robot has to "eat" and "work" at least. In section 3, a criterion is described to judge the performance of an agent over several cycles of iterated "work" and "eating". This criterion works fine in experiments to learn the basic behaviors of the agents in the ecosystem in simulations. Section 4 discusses the criterion used in the previous section in some detail. Its relation to self-sufficiency is explained and we argue why it can be useful for learning in autonomous robots in general. In addition, we show why the criterion fails in the real-world under certain circumstances. As we see in section 5, the problems are related to the fact that the criterion does a long-term analysis. We argue that immediate and strong feedback, similar to pain, is needed as well. Accordingly, an additional criterion is presented which monitors sudden changes in the power-consumption. Section 6 presents a learning algorithm based on the combination of both criteria. In section 7 concrete experimental results are presented. Section 8 concludes the paper.

2 Robots in an ecosystem

In this section, we present a special set-up, the so-called ecosystem, where the notion of "try to stay alive" has a very concrete meaning. The ecosystem (figure 1) originates in work of Luc Steels [Ste94b] with a biological background motivated by David McFarland [McF94]. The basic set-up consists of autonomous robots, competitors and a charging station. The robots are LEGO vehicles (figure 2) with two motors, bumpers, active InfraRed, and sensors for white light, modulated light, voltage, and current. The robots are autonomous as they are equipped with a carried-on microcontroller-board and re-chargeable batteries. Some of the experiments described lateron were done with a new more "professional" robot-base developed recently in our lab. The main difference is that this new robot operates much more reliable over longer periods. But in principle, i.e., in respect to the sensors and effectors used in the experiments described here, the LEGO vehicles and the new base are identical.

Fig. 1. The ecosystem with (from left), a competitor, the charging station, and an autonomous robot.

The competitors in the ecosystem are lamps connected to the same global energy-source as the charging station. Therefore, they take valuable resources away from the robots. But the robots can defend themselves. If a robot pushes against a box housing a lamp, the light of the lamp dims. As a consequence, there is more energy available for the robots. But only for a restricted time as the dimming of the lamp slowly decreases.

The following "tasks" are of basic importance for an agent in the above described environment:

Recharge : To avoid "death", i.e., not to run out of energy, the agent has to recharge from time to time its batteries.

Fig. 2. The robot build with LEGO^{TM}.

Obstacle Avoidance : To prevent physical damage, the robot has to react to obstacles and to turn away from them.

Fight Parasites : To get more energy — the environmental settings can be such that the robot dies without dimming the lamps —, the robot has to push against the boxes with the lamps.

When following a behavior-oriented approach, one is not obliged to program explicit functions for solving these "tasks". Instead, as demonstrated in some detail in [Ste94b], it is sufficient to use simple basic behaviors and to let the complex functionality emerge from interactions with the environment.

The "fighting" as pushing against boxes is for example caused by two elementary behaviors: touch-based obstacle avoidance and attraction to modulated light. The attraction to light leads the robot to drive towards a box housing a lamp, it will bump into it, retract due to obstacle avoidance, being attracted again, and so on until the light is dimmed and the attraction is therefore gone.

We claim that this set-up can be used to gain some general insight about self-sufficient robotic agents. Alike Masanao Toda's "Fungus Eaters" [Tod82], our robot's have "a job to do". The pushing of the competitors is a kind of *work* which is paid in terms of energy. The interesting aspect of our set-up is, that we can understand and even control the underlying dynamics and parameters in respect to working (among others). It follows, that we can cover many different scenarios; in some of them a certain way of "doing your job" is efficient, whereas in others the same strategy fails. For example, for two robots it is beneficial to cooperate in fighting for some settings, whereas with other settings it is better for the robots to work on their own. Criteria that guide learning in all these scenarios can thus be claimed to have some generality.

3 Learning experiments in simulations

An exciting challenge is to learn a good behavioral control of a robot in the above described eco-system. Steels reports in [Ste96a] and [Ste96b] successful experiments in simulations. In these experiments a robots is controlled by so-called *couplings* between quantities. Quantities are bounded variables of system parameters like e.g. speed-of-the-left-motor or right-bumper. A coupling between two quantities a and b establishes an "influence" of b on a. The overall control of the robot is determined by a string of couplings, called *bene* (behavioral gene). The benes are learned with a selectionist mechanism. In the simulated experiments following basic behaviors are learned: obstacle-avoidance, photo-taxis towards the charging-station, stopping in the station while charging, and photo-taxis towards the competitors (resulting in "fighting" them).

Of our main interest in this paper is the selection criterion used in these experiments, which is based on the notion of *viability* [Ash52]. As mentioned in the introduction, a behavior is seen as viable when it keeps the organism within a region of the parameter space that guarantees continued existence [MH81]. As the energy-level is one of the most important parameters for a robotic agent, the performance of a bene during time interval T is evaluated via the *min/max-level(T)*

$$min/max\text{-}level(T) = min(T) + (max(T) - min(T))/2$$

with $min(T)$ the minimum and $max(T)$ the maximum energy level of the agent during time period T. A good bene is expected to raise the *min/max-level*.

Unfortunately, Steels denoted this criterion as *center of mass* which is somehow misleading. The *min/max-level* is neither an average of the energy-level over time nor its center of mass in the notion from physics. This is illustrated in figure 3. There, we can see that this criterion depends only on the minimum and the maximum energy-level during time-period T, without taking the slope of the function into account. In situation A, a criterion using an average or the actual center of mass would lead to a higher value than the *min/max-level*. In situation B, it would lead to a smaller value. As we will see, we believe it to be important, that the *min/max-level* is the same in both situations.

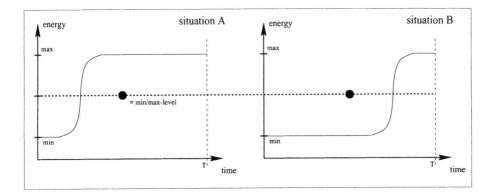

Fig. 3. Though situation A and B lead to different averages over time and centers of mass of the energy-level, the *min/max-level* is in both situations the same.

4 The *min/max-level* in the Real-World

As mentioned before, the *min/max-level* has been used (though denoted differently) by Steels to learn successfully basic behaviors in the ecosystem in simulations. We explain here, why we believe this criterion to have some general importance for self-sufficient robots. In addition, we will indicate problems when this criterion is carried over to the real-world and how these can be overcome.

The *min/max-level* is meant to be a criterion which judges the long-term performance of a robotic agent. A self-sufficient agent has to fulfill a task. In doing so, we can expect him to use up energy. As an autonomous robot can be expected to run on batteries, there has to be a time when the robot has to refill them again. Therefore, the behavior of a self-sufficient robot can be seen in general as cycles switching between "work" and "feeding". The time-period T over which the *min/max-level* is monitored should be such, that it includes at least one such cycle. Then, the minimum corresponds to the state where the robot has used up his internal energy-resources the most (while working), and the maximum corresponds to the state where the robot has filled its batteries the most (while charging). When the *min/max-level* moves up, there is some indication that the agents did his job more efficiently as he "exhausted less" (the consumption of energy was less) and/or he got "more work done" (the pay-off in form of energy was higher).

From the above discussion, the disadvantages of the *min/max-level* are almost immediately clear when it comes to the real world:

1. The *min/max-level* is not a meaningful criteria if the robot does not "eat" during the time period T.
2. Learning always involves a "trial" component. This can lead to dangerous situations which have to be handled fast. The feedback provided by the *min/max-level* is then too slow.

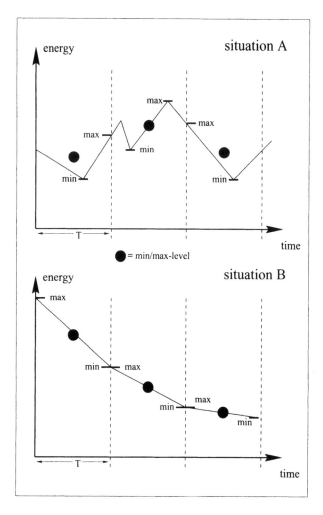

Fig. 4. The *min/max-level* with and without "eating", i.e., increasing the energy level by recharging.

First, let us have a closer look on problem One which is illustrated in figure 4. The *min/max-level* criteria works fine if the robot "eats" at least once during the time period T, i.e., it does occasionally increase its energy-level by re-charging (situation A). If not, the *min/max-level* continuously decreases (situation B). Therefore, no change in the robot's behavior will reveal that it is better than the previous one.

In the experiments to learn basic behaviors in simulations the problem does not occur. The robot starts with a random walk in which it rarely eats, but it happens. Especially, it does not starve. His energy level oscillates somewhere between 40 and 75 percent of its maximum capacity. This is infeasible to map to

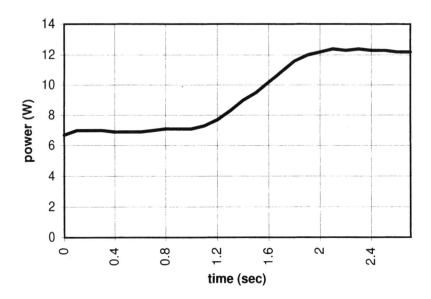

Fig. 5. The rise in the power-consumption when the robot is blocked by an obstacle.

the real-world. If we start a random walk, the robot very likely gets stuck at a wall. Even if we provide him with obstacle-avoidance, the chances of driving in the charging-station, let alone picking up a non-neglectable charge, are incredibly low.

Let us now have a closer look to the second problem, the "slowness" of the *min/max-level*, which we illustrate with some concrete results from experiments. When a robot drives straight without being hindered, it has a power-consumption of roughly 6.4 W. When hitting an obstacle which prevents further movement, the wheels stall and the current in the motors rises dramatically. The power consumption rises as a result to roughly 12.2 W. Figure 5 shows this situation; the robot is stopped [1] by an obstacle at 1.2 seconds.

When the batteries are full, they provide roughly 13 Wh. Suppose, we measure the *min/max-level* with $T = 5$ minutes of a robot starting with roughly fresh batteries. Given the following two situations of

1. driving un-hindered and
2. being **continuously** stopped by an obstacle

then the *min/max-level* will be in the first case at 12.74 Wh and in the second

---

[1] The power rises not straight but rather smoothly in this graph. This is mainly due to the fact that we use some averaging-functions in the robots to get rid of noise.

one at 12.49 Wh. Though the motors can be seriously harmed during this period, the "indication" that driving un-hindered is much better is rather low.

Let us view the problems with the notion of viability in mind. The criterion of the *min/max-level* works fine in a "halfway-safe" situations, i.e., when the survival of the agent is guaranteed (for some time) and the task is to make his performance more efficient. But it fails when the essential parameters come very close to the borders of the viability zone or simply "dangerous" situations. Then, fast and significant feedback is needed. The *min/max-level* gives some long-term feedback over the status of viability, loosely comparable to hunger. What we need in addition to make this criterion useful in the real-world is a mechanism comparable to pain.

5   Energy-consumption as a fast feedback

Before we introduce the "pain-like" criterion which allows a fast feedback for the learning in our real-world experiments, we have a closer look on the way we build behaviors. As in the simulated experiments by Steels, we use a selectionist mechanism for generating new encodings of behaviors, the so-called *benes* (see section 3). As basic building-blocks we use *sensor-activations*, which calculate a value $a \in [0;1]$, and *elementary behaviors* which are controlled by an activation $a \in [0;1]^2$. A coupling between a sensor-activation and an elementary behavior simply feeds the output of the sensor-activation into the activation value of the elementary behavior. A bene is a set of couplings.

Following sensor-activation are used in our experiments:

bump-left is "triggered", i.e., its value rises to One, if the left bumper is touched and decays within some seconds to Zero as the contact is gone.
bump-right analogous
top-bump is directly linked to the charging-antenna, i.e., the output is One when the robot is in the charging-station, Zero else.
IR-left outputs a scaled amount of how much more the left active IR sensor is activated than the right one.
IR-right analogous
white-light-left outputs a scaled amount of how much more the left white light sensor is activated than the right one.
white-light-right analogous

As elementary behaviors that can be coupled with the above sensor-activations we use:

retract-left: move back and turn slightly to the left
retract-right: analogous
stop: halt the robot
turn-left: add some left turn to the robots motion

---

[2] This is similar to an approach proposed in [Jae96] where differential equations are used to model so-called activation- and target-dynamics of behaviors.

turn-right: analogous

When starting with nothing but a roam-behavior, i.e., a forward-movement with a speed of 80%, we want to learn following more complex basic-behaviors:

touch-based obstacle avoidance
    with the couplings
       – *bump-left* ⇒ *retract-right*,
       – *bump-right* ⇒ *retract-left*
active IR obstacle avoidance
    with the couplings
       – *IR-left* ⇒ *turn-right*,
       – *IR-right* ⇒ *turn-left*
re-charging
    with the couplings
       – *white-light-left* ⇒ *turn-left*,
       – *white-light-right* ⇒ *turn-right*,
       – *top-bump* ⇒ *stop*

For all of these couplings, it can be shown that they are impossible or at least hardly learnable with the *min/max-level* in the real-world set-up. This is mainly due to the problem, that this criterion fails if no charging occurs. This is not surprising, as it is suited as pointed out before mainly for learning more efficient behaviors. Here, we actually need to learn most basic behaviors.

Therefore, we need a criterion giving fast and strong feedback to help the agent to "get going" first (plus to help the agent to handle "emergencies"). We propose to use the (short-term) energy-consumption as an indicator for this purpose. Let us for example have a closer look on how touch-based obstacle avoidance is learned with this criterion. In doing so, we ignore the concrete time intervals used for measuring the energy-consumption in the rest of this section. Therefore, we only look at the power for reasons of convenience.

As mentioned before, the power consumption increases dramatically when the robot hits an obstacle (figure 5). In addition, the robot gets stuck in that situation. When it tries different couplings, most of them do not change the situation at all. A few ones make the situation even worse. Figure 6 shows a situation where at timepoint 4.0 seconds a coupling is used that increases — or actually tries to increase — the motor speed. For example feeding the output of the active InfraRed, which is high when the robot is pushing against an obstacle, into one of the turn-processes, creates such a bad coupling. Only when one of the bumper (or active IR) outputs is coupled to a retract behavior, the power-consumption goes down.

6 Combining the fast feedback with the min/max-level

Using only the criterion of minimizing the (short-term) energy-consumption has a severe flaw: the robot has the smallest energy consumption when it does not

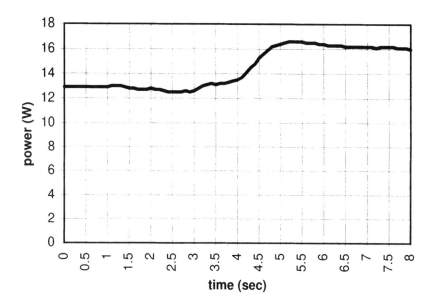

Fig. 6. Trying a coupling that increases motor activations while being stopped by an obstacle.

move at all. When using only this criterion, the robot discovers that the best thing to do is to sit and wait (when e.g. coupling white-light perception to stop while being close to the chargings station). After a while, it happily dies. But when using the criterion of the min/max-level the "sit and wait" can be detected as being "bad" as it prevents recharging.

Furthermore, the fast feedback is most suited to evaluate *single couplings* whereas the min/max-level is intented for *whole benes*, i.e., sets of couplings. For example the discovery that the coupling *top-bump* ⇒ *stop* (stop in the charging-station to fill batteries) is useful is only possible if the robot is using phototaxis at the same time, i.e., while evaluating a whole bene.

Therefore, we combined both criteria in a learning process working in two modes. The default mode uses the min/max-level. The second "emergency"-mode is triggered by an abrupt rise in the power-consumption and works with the fast feedback. Concretely, the learning algorithm starts with an empty bene $b_0 = \{\}$ and a min/max-level of 0 in mode 1 (figure 7). In case an increase in the power-consumption of more than 3% is detected, mode 1 is aborted and the system starts mode 2 (figure 8). Note that mode 1 evaluates whole benes whereas mode 2 only uses single couplings.

```
Mode 1 (default): given bene b

randomly generate a coupling c
if c ∉ b
      generate b' by including c in b
else
      generate b' by excluding c in b
run robot with b' for 30 seconds
if max-level does not rise
      reject b'
      start again with b
else
      use b' for 4.5 more minutes
      if the min/max-level rises
            keep b' (discard b)
            start again with b'
      else
            reject b'
            start again with b
```

Fig. 7. pseudo-code of mode 1

```
Mode 2 (emergency):

do
      randomly generate a coupling c
      run robot with c for 0.5 sec
until power-consumption drops
add c to b and b'
start Mode 1 again
```

Fig. 8. pseudo-code of mode 2

7 Results

Let us illustrate the working principle of the algorithm from the previous section
with some concrete results from an actual run (figure 9). Starting with an empty
bene b0 in mode 1, the coupling *white-light-left* $\Rightarrow$ *retract-right* is randomly
generated. The coupling is added to the empty bene b0 leading to the bene b1
which is evaluated. The current maximum and minimum energy-level as well as
the actual $(max - min)/2$ are printed in 5 second intervals. The min/max-level
constantly drops during the 30 seconds evaluation time (leading to no further

evaluation and rejection of b1) as this bene with its single non-sense coupling does not lead to any useful behavior.

In addition, an emergency occurs (the robot with its non-sense bene hits the wall and gets stuck) during the evaluation of b1 at timepoint 13 seconds. This causes mode 2 to be activated and different randomly generated couplings are tried for 1 second each. The coupling *bump-left* ⇒ *retract-left* is discovered as being usefull (as it frees the robot) and added to both benes b0 and b1. After this interruption for 4 seconds, the evalution in mode 1 of the (now altered) bene b1 is continued. As mentioned above, the b1 is rejected after 30 seconds. Mode 1 starts again with the (now altered) bene b0, a new coupling *IR-right* ⇒ *retract-right* is generated, then added to b0 to generate a new b1, and so on.

In this manner several couplings which avoid the "getting stuck" are discovered during the first 3 minutes. Some of them lead to a somewhat "strange" form of obstacle-avoidance: when the robot couples *bump-left* to *retract-left* (instead of *retract-right*) while hitting an obstacle (like it did in its first "discovery"), the robot moves away from the obstacle and thus the power-consumption falls. But the robot turns in the wrong direction and therefore it it is likely to bump (though with a different angle) into the obstacle again. Nevertheless, after a few "unnecessary" bumps the robot gets away.

Something really interesting happened again after 11.5 minutes. The robot learns (partial) phototaxis (see figure 10). The existing bene b0 (with some couplings for obstacle-avoidance) is augmented with the coupling *white-light-left* ⇒ *turn-left* which leads to "one-sided" phototaxis, i.e., the robot is attracted to the station if he approaches from the left side. This coupling is sufficient to cause an entering of the station (somewhere between 705 and 710 sec). While crossing the station some charge is picked-up and the minimum energy-level stops falling while the maximum energy-level even slightly rises. Therefore, evaluation of the bene is continued for further 4.5 minutes. During this time 4 additional "hits" of the station occured which lead in the end to a significant rise in the min/max-level. Therefore, bene b1 is selected.

The coupling *top-bump* ⇒ *stop* is learned 8 minutes later in the same manner. The re-charging behaviors are complete with the learning of *white-light-right* ⇒ *turn-right* another 2.5 minutes later. Interesting enough, the "smooth" obstacle-avoidance with active IR learned quite late in this particular run. The two according couplings were present in the robot's bene after 27.5 minutes total run-time in this particular experiment.

8 Conclusion

We presented in this paper two learning criteria based on the idea that the energy-level is a crucial factor for the viability of autonomous robots. The first criterion monitors long-term developments similar to ideas of McFarland and Spier. Results are presented from simulations done by Luc Steels where this criteria is used to learn the basic behaviors of a robot in an ecosystem. We argue

```
************************************
b0={}
MODE 1
b1={wll*rer}

t MAX MIN mml-diff
0 .0 .0 .0
5 .0 -3.2 1.6
10 .0 -6.4 3.2

************************************
EMERGENCY: enter MODE 2 (t=13)
t coupling power
14 tob*sto 12.4 -> 12.4
15 wll*rer 12.4 -> 12.4
16 irl*tul 12.4 -> 16.3
17 bul*rel 12.4 -> 6.7
************************************
b0={bul*rel}
b1={wll*rer, bul*rel}
re-enter MODE 1
************************************

t MAX MIN mml-diff
20 .0 -15.1 7.5
25 .0 -21.7 10.8
30 .0 -28.3 14.1

mml: previous = .0 / final = -14.2
=> reject b1
************************************

b0={bul*rel}
MODE 1
b1={bul*rel, irr*rer}

t MAX MIN mml-diff
30 -28.3 -28.3 .0

and so on...

Annotations:
bul = bump-left, bur = bump-right, tob = top-bump, irl = IR-left, irr = IR-right, wll = white-light-
left, wlr = white-light-right, rel = retract-left, rer = retract-right, tul = turn-left, tur = turn-right,
sto = stop
```

Fig. 9. The start of the learning-process: discovering (part of) the touch-based obstacle avoidance.

```
...and so on

**************************************
b0={bul*rel,irl*rel,bur*rel}
MODE 1
b1={bul*rel,irl*rel,bur*rel,wll*tul}

t MAX MIN mml-diff
690 -467.8 -467.8 .0
695 -467.8 -471.0 1.6
700 -467.8 -474.2 3.2
705 -467.8 -477.4 4.7
710 -467.8 -477.4 4.7
715 -459.0 -477.4 9.2
720 -449.8 -477.4 13.8


**************************************
RISE in MAX-LEVEL: continue evaluation
**************************************

725 -449.8  -477.4  13.8
730 -449.8  -477.4  13.8
735 -449.8  -477.4  13.8

...and so on...

980 -407.3 -477.4 35.0
985 -407.3 -477.4 35.0
990 -407.3 -477.4 35.0

mml: previous = -453.6 / final = -442.4
=> keep b1 (reject b0)

and so on...
```

Fig. 10. Learning "one-sided" photo-taxis.

that the name used by Steels for this criterion (center of mass) is misleading and propose a new one (min/max-level).

It is shown that the results from the simulations do not carry over to the real-world set-up. This is explained including a discussion and motivation of the min/max-level criterion from the viewpoint of self-sufficiency. We argue, that this criterion is suited to measure how efficiently a robot "does his job". The min/max-level monitors over cycles of iterated "work" and "recovery". In doing so it indicates if the robot "exhausted less" (the consumption of energy was less) and/or if it got "more work done" (the pay-off in form of energy was higher) during a fixed time period compared to previous ones.

Due to its long-term character, the min/max-level fails in two kinds of situations. First, it fails when learning from scratch, i.e., when the most basic behaviors have to be found. In this situation, "finding food" is very unlikely which renders the min/max-level a useless criterion. Therefore, a faster mechanism is needed then for "getting going". Second, the min/max-level fails when the robot encounters dangerous situations, i.e., the border-lines of its viability-zone. Such situations are somehow likely to occur in general as any learning mechanism has a certain component of trial and error. Obviously, these situations are of greater importance in the real-world (were they can lead to expensive damage) than in simulations (were they sometimes are simply "modeled away").

We claim that both kinds of situations should be handled by a fast and strong feedback similar to pain (whereas the min/max-level could be somehow compared to hunger). We propose to monitor sudden changes in the energy-consumption for this purpose. Experiments in the real-world set-up are presented were a robot learns the most basic behaviors using an algorithm which combines this criterion and the min/max-level. Namely, the robot learns so-called couplings between simple sensor-data and effector-control which lead in combination to touch-based as well as active IR obstacle avoidance, and to recharging.

9 Acknowledgments

Many thanks to all members of the VUB AI-lab who work jointly on maintenance and improvements of the robots and the ecosystem, as well as on the concepts behind it. The robotic agents group of the VUB AI-lab is partially financed by the Belgian Federal government FKFO project on emergent functionality (NFWO contract nr. G.0014.95) and the IUAP project (nr. 20) CONSTRUCT.

References

[Ash52] Ross Ashby. *Design for a brain*. Chapman and Hall, London, 1952.

[Jae96] Herbert Jaeger. The dual dynamics design scheme for behavior-based robots: a tutorial. Technical Report 966, GMD, St. Augustin, 1996.

[McF94] David McFarland. Towards robot cooperation. In Dave Cliff, Philip Husbands, Jean-Arcady Meyer, and Stewart W. Wilson, editors, *From Animals to Animats 3. Proc. of the Third International Conference on Simulation of Adaptive Behavior*. The MIT Press/Bradford Books, Cambridge, 1994.

[MH81] D. McFarland and A. Houston. *Quantitative Ethology: the state-space approach*. Pitman Books, London, 1981.

[Min85] Marvin Minsky. *The Society of Mind*. Simon and Schuster, New York, 1985.

[MS97] David McFarland and Emmet Spier. Basic cycles, utility and opportunism in self-sufficient robots. *Robotics and Autonomous Systems (in press)*, 1997.

[Pfe96] Rolf Pfeifer. Building "fungus eaters": Design principles of autonomous agents. In *From Animals to Animats. Proc. of the Fourth International Conference on Simulation of Adaptive Behavior*. The MIT Press/Bradford Books, Cambridge, 1996.

[Sim83] Herbert H. Simon. Why should machines learn? In Tom M. Mitchell Jaime G. Carbonell, Ryszard S. Michalski, editor, *Machine Learning: An Artificial Intelligence Approach.* Tioga, Palo Alto, 1983.

[Ste94a] Luc Steels. The artificial life roots of artificial intelligence. *Artificial Life Journal, Vol 1,1*, 1994.

[Ste94b] Luc Steels. A case study in the behavior-oriented design of autonomous agents. In Dave Cliff, Philip Husbands, Jean-Arcady Meyer, and Stewart W. Wilson, editors, *From Animals to Animats 3. Proc. of the Third International Conference on Simulation of Adaptive Behavior.* The MIT Press/Bradford Books, Cambridge, 1994.

[Ste96a] Luc Steels. Discovering the competitors. *Journal of Adaptive Behavior 4(2)*, 1996.

[Ste96b] Luc Steels. A selectionist mechanism for autonomous behavior acquisition. *Journal of Robotics and Autonomous Systems 16*, 1996.

[Tod82] Masanao Toda. *Man, robot, and society.* The Hague, Nijhoff, 1982.

[Wil91] S.W. Wilson. The animat path to ai. In *From Animals to Animats. Proc. of the First International Conference on Simulation of Adaptive Behavior.* The MIT Press/Bradford Books, Cambridge, 1991.

Perceptual grounding in robots

Paul Vogt

Vrije Universiteit Brussel, Artificial Intelligence Laboratory
Pleinlaan 2
1000 Brussels, Belgium
e-mail: paul@arti.vub.ac.be

Abstract. This paper reports on an experiment in which robotic agents are able to ground objects in their environment using low-level sensors. The reported experiment is part of a larger experiment, in which autonomous agents ground an adaptive language through self-organization. Grounding is achieved by the implementation of the hypothesis that meaning can be created using mechanisms like feature generation and self-organization. The experiments were carried out to investigate how agents may construct features in order to learn to discriminate objects from each other. Meaning is formed to give semantic value to the language, which is also created by the agents in the same experiments. From the experimental results we can conclude that the robots are able to ground meaning in this self-organizing manner. This paper focuses on the meaning creation and will only discuss the language formation very briefly. The paper sketches the tested hypothesis, the experimental set-up and experimental results.

1 Introduction

In the past few decades a lot of research has been done on the grounding problem. The grounding problem investigates how symbolic structures may arise from non-symbolic sensory information extracted from the physical world (Harnad, 1990). This paper discusses how robotic agents may ground symbolic meaning from sensory observations. Approaches to this problem normally use rule-based systems, connectionist models or genetic learning mechanisms. Our approach is a bottom-up approach, conform the behavior-oriented approach to artificial intelligence (Steels, 1994). This means that agents use sensory information to conceptualize meaning from the bottom-up, which has also been suggested by Harnad. The experiments are embedded in our research on the origins of language (Steels, 1996a).

The experiments are based on the hypothesis that language is a complex dynamical system and that a lexicon and meaning emerges through interaction within a population of agents and their environment (Steels, 1997). Cultural interaction between agents causes the language to spread through the population. Individual adaptation lets the lexical and semantic symbols be categorized. Because the system is a distributed complex dynamical system, there is no agent that has a complete view of the language, nor is there an agent that can control the language. The major goal in studying the origins of complex systems is to find the boundary conditions and mechanisms, like self-organization, that give rise to the complexity, see e.g. (Prigogine and Strengers, 1984).

In our research on the origins of language, see e.g. (Steels, 1996c), agents can adapt a *lexicon*, which are words that they associate with semantic symbols, which in turn discriminate between concepts (be it objects or other relations). The lexicon is propagated through the population of agents by means of social interaction. This is done by a distributed set of agents that have the ability to communicate with each other, and a self-organizing learning method is used to adapt the lexicon appropriately. No linguistic information is pre-programmed and the lexicon emerges through a cultural evolution. This is opposed to the linguistic theory introduced by Chomsky (see e.g. (Chomsky, 1980)), who claims that linguistic structures are innate.

As there is no linguistic information pre-programmed there is also no pre-programmed meaning. In the experiments individual agents ground the semantic symbols. The robots ground the symbols, which compose the linguistic meaning of a concept, i.e. the meaning of a word. They do so by classifying sub-symbolic low-level sensory data into distinctions, conform (Harnad, 1990). The grounded symbols (or feature sets) distinguish one object from some other observed objects. An adaptive self-organizing learning mechanism is incorporated in the system in order to select those symbols that are useful for the lexicon formation (Steels, 1996b). Learning is achieved by generating features, which relate to observed sensory information of an object. These features are then classified based on their usefulness by means of a selectionist approach. This method is based on natural selection during the interaction with the physical world and within a lifetime (as suggested by e.g. (Foley, 1997)(Edelman, 1987)). The useful features will be used more frequently than unuseful ones. The useful features remain within the system, whereas the unuseful ones will be forgotten. The method differs from genetic algorithms because the agents do not inherit information from generation to generation. All learning and selection takes place within a lifetime of an agent. They also do not reproduce features for selection, but they create new features.

This paper reports on an experiment where robotic agents can develop a lexicon about objects in their environment using these mechanisms. In (Steels and Vogt, 1997) a similar experiment is reported, in that paper the focus was however on the co-evolution of language and meaning, conform (Steels, 1997). Here the focus is on the evolution of meaning in robotic agents. Especially the physical implementation of the hypothesis is discussed in detail.

In the next section the mechanisms for meaning creation and learning are explained. In section 3 the whole experimental set-up is described. Section 4 reports on the experimental results. And finally, in section 5, conclusions and future research are discussed.

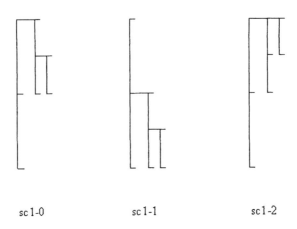

sc 1-0   sc 1-1   sc 1-2

Fig. 1. The robots can construct a binary tree of features that divides the sensor space.

2 The evolution of meaning

At the AI-Lab in Brussels we investigate how meaning may evolve using se-lectionistic mechanisms (Steels, 1996b). These mechanisms are mainly based on adaptive self-organization. The mechanisms for the meaning creation are (1) *fea-ture generation* and (2) *self-organization* . The generation principle is based on the generation of features that may represent sensory information of an object. Self-organization by means of a natural selection-like learning method causes the meaning to be coherent and useful.

The meaning creation is introduced by Luc Steels (Steels, 1996b). The princi-ples for meaning creation are valid for all kinds of concepts like physical objects, spatial relations, internal states and movements. It emerges in an individual agent under the notion of *discrimination games*: Agents construct binary fea-ture trees of the sensory space representing objects (figure 1). Observed objects are related with feature nodes that are sensitive for observed sensor values. These features are used to discriminate one object from other objects. Those features that are capable to distinguish one object from another constitute the feature sets and represent the symbolic meaning. The following description is adapted from (Steels, 1996b).

Suppose we have a system that consists of a set of objects $O = \{o_1, \ldots, o_n\}$ and a set of agents $R = \{r_1, \ldots, r_m\} \subset O$. Each agent r has a set of sensory channels $S_r = \{sc_1, \ldots, sc_p\}$ and a set of features $F_r = \{f_1, \ldots, f_q\}$. A sensory channel sc_j is a real-valued function from $O \to \Re$, which is (in-)directly derived from the agent's sensor. A feature f is a pair $(p\ V)$, where p is the attribute name (usually sc_j) and $V = [v_{min}, v_{max}] \in SC_j$. Here SC_j is the range in which sc_j is sensitive.

In a discrimination game, the procedure is as follows: Suppose an agent has observed a context that consists of a set of objects: $C = \{o_i \mid o_i \in O\}$. Observing an object in the robot's world using (low-level) sensors results in a set of sensor values for every sensory channel. These values can be related to features that are sensitive to these values. So every object $o_i \in C$ can be related to a set of features $F_{r,o_i} \in F_r$. An object o_i can be described by a set of *descriptions*:

$$\Lambda^{o_i} = \{\Lambda_k^{o_i} \mid \Lambda_k^{o_i} \subseteq F_{r,o_i}\}$$

where $\Lambda_k^{o_i}$ is any possible description of an object o_i constrained as follows:

$$\Lambda_k^{o_i} \subseteq \{f_j = (p_j V_j) \mid f_j \in F_{r,o_i} \wedge (l \neq j \Rightarrow p_l \neq p_j)\}$$

Note that: (1) The set on the right hand side of this relation may result in several sets since the o_i may be related to several features on one sensory channel, since the features are build up in a tree. The relation holds for every possible set governed by the right hand side. And (2) $\bigcup_k \Lambda_k^{o_i} = F_{r,o_i}$.

The agent now identifies a *topic* $t \in C$ of the discrimination game. Then the agent tries to distinguish the topic from the other objects in the context by constructing a *set of distinctive feature sets*. A set of distinctive feature set can be defined as:

$$D_t^C \equiv \{\Lambda_j^t \mid \forall o_i \in C \setminus \{t\} : (\neg \exists \Lambda_k^{o_i} : (\Lambda_k^{o_i} = \Lambda_j^t))\}$$

So, the agent constructs a set of features that relates to t and which distinguishes t from the other objects in the context. A distinctive feature set is associated with a use-score and a success-score. A discrimination game results in a *set of distinctive feature sets* (D_t), which can be empty, but which can also consist of several distinctive feature sets. All distinctive feature sets resulting from the discrimination game are stored in the memory of the robot to be associated words in the language games.

If the agent thus determined a set of distinctive feature sets, the following is possible:

- The agent was not able to discriminate $t \in C$, i.e. $D_t = \emptyset$. The discrimination game was a failure.
- The agent was able to discriminate $t \in C$, i.e. $D_t \neq \emptyset$. The discrimination game was a success. If there is more than one distinctive feature set, the agent will choose the most general distinctive feature set as follows:
  - Choose the set with the smallest amount of features.
  - If there is still more than one distinctive feature set, then choose the set that is least segmented in the feature space. I.e. that set that has the least average depth in the feature tree.
  - If there are still several sets left, then choose the most successful one. I.e. the set for which the measure $m = success/use$ is the highest. (Steels, 1996b).

If the discrimination game was a failure, then the agent constructs a new feature. This can be done in two ways: (1) If there is still a sensory channel unused, then use this sensory channel to construct a feature in the range of this channel. (2) Choose an arbitrary feature that is 'activated' by the topic, and

Fig. 2. The LEGO robots used at the AI-Lab in Brussels.

divide the range of this feature in two equal halves, thus constructing two new features. This way a binary tree of features is created (figure 1).

If the discrimination game was a success, the use- and success-scores are updated as follows: The use parameter is incremented for all distinctive feature sets. The success-score is incremented for the distinctive feature set, which represents the object the best. Self-organization is achieved by the way the agent selects a distinctive feature set in using these scores. So, learning to ground the meaning of objects is achieved by generating perceptual features, then trying to distinguish an object from others and increasing the strength of the feature sets that were successful in the discrimination task.

3 The experiments

3.1 The robots and their environment

At the AI-Lab in Brussels we have several mobile LEGO robots (figure 2) which can explore an ecosystem of approximately 4 by 6 meters (see figure 3). On the robots several sensor modules are mounted for sensing and two motors to drive them. The process unit of the robots is a Motorola MC68332 micro controller with 128 kB ROM and 256 KB RAM located on a Vesta board. The Vesta board is integrated with the SMB2 sensory-motor board, which is dedicated to low-level sensory-motor processing (Vereertbrugghen, 1996). The robots have two white light and two modulated light sensors, each on the left and the right front side of the robot. Three infrared (IR) sensors are mounted on the left-, middle- and right front side. Four IR emitters are mounted on the robot in such a way that they emit IR in four perpendicular directions (to the front, the back, the right

Fig. 3. The ecosystem of the Al-Lab in Brussels. In the foreground we see a competitor that emits modulated light, and in the background the charging station emitting white light is visible.

and the left sides). Four bumpers are used for obstacle avoidance. Secondary NiMH batteries supply the power (Birk, 1997).

Communication between the robots is done by a radio-link that can transmit and receive messages up to 40 Kbit/s. This radio-link is part of the sensory-motor board, and is therefore controlled by the processor of this board. The radio-link is unreliable in the sense that it is not guaranteed that a message arrives, but when it arrives the message contains no errors (Vereertbrugghen, 1996). The radio-link is used for both linguistic and extra-linguistic communication.

The robots are programmed in PDL, a language that is designed for behavior-oriented control as described in (Steels, 1994). The system directly couples sensory data streams to the actuators, thus controlling behavior dynamically. Behavior is implemented as a motivational coupling between sensory input and motor output.

In the robot's environment (see figure 3) there are several objects for which the robots need to ground a semantic representation. These objects can be described by the light source they emit. There is one light source that emits white light, there are two light sources that emit light modulated at a particular frequency, and the robots emit infrared to make themselves visible for the other robot. The whole system, i.e. robots and environment, is used for the meaning and language experiments described in this paper. In the next section this experiment is described in more detail.

3.2 The language experiments

In order to investigate the origins of language on physical robots we constructed an experiment in which robots engage in a series of *language games*. At first the robots have no knowledge of language at all; they only have knowledge about how to communicate (Steels and Vogt, 1997). In the experiment described here, there are only two robots that play a series of *naming games*. In naming games (a variant of the language games) the agents indicate objects by their name (Steels, 1996a). The objects about which the robots can communicate are, apart from the robots themselves, an object that emits white light and two objects that emit light modulated at a particular frequency.

In a language game, one robot decides to be the speaker, while the other becomes the hearer. In order to construct a coherent context, both robots must know what objects are present in their near surroundings and where these objects are. This can be done while the robots are standing close to each other and facing one another. So, when the robots are facing each other at close distance they have to build a map of their surroundings, thus constructing a context of objects. The speaker chooses one object randomly from the context as the topic. Then the speaker uses extra-linguistic communication (as described later) in order to let the hearer identify this topic as well. For this object each agent constructs a set of distinctive feature sets in a discrimination game as described in section 2. Now that both agents have a set of distinctive feature sets they can start the communication as described in more detail in (Steels and Vogt, 1997)

The experiment is implemented so that the robots can be in three different modes: (1) default exploration mode, (2) the speaker mode, and (3) the hearer mode. In the default exploration mode, the robots drive around in the ecosystem in forward direction with touch-based obstacle avoidance. Each robot emits pulses of IR, so that they can detect the presence of another robot. When no IR is emitted, but the robot senses IR, it can infer that there must be another robot nearby. The robot that detects the other then broadcasts a radio message that it wants to communicate. The other agent then replies a confirmation to the first one, and enters the hearer mode. When the first robot receives the confirmation, it enters the speaker mode. This type of communication is extra-linguistic, so it does not add to the language being formed.

The hearer starts to emit a continuous signal of IR, and the speaker uses IR-taxis (or phototaxis towards infrared) to drive towards the hearer as described by (Braitenberg, 1984). The motor values are adjusted so the robot drives towards the IR source. When the speaker is close to the hearer, it stops and uses *IR-orientation* to orient towards the hearer. With IR-orientation, as with IR-taxis, the robot orients towards the highest gradient of difference between the outer IR sensors, but without forward drive. After the speaker has completed this, the hearer orients towards the speaker, so they now stand in front of each other.

One by one the speaker and the hearer start to scan their surroundings. They do so by rotating around their axis while recording their sensory input, thus constructing a map like in figure 4. As we can see in figure 4(a), the complementary sensor pairs show intersections at, for instance, time/position 100, 140 and 160. The first one is an intersection of modulated light; the second one is from white light. IR intersects at the beginning and the end of the graph. The intersections occur when the robot passes an object, so this information could be used to decide that an object is at that particular place. If such an intersection is detected, the robot makes a record of all sensory channels. These values may be direct sensor values or real valued functions of direct sensor values. The construction of such sensory channels is very important for the success of the discrimination games. This construction will be discussed in the next section.

If the speaker has chosen a topic, it has to use extra-linguistic means in order to let the hearer identify the topic as well. We have decided to use pointing as method for this process. Pointing is implemented as follows: The speaker orients towards the topic, while it emits IR in four perpendicular directions. The hearer observes the pointing and perceives a graph as shown in figure 5. In this figure, one can observe three relative high peaks of IR. Each peak is at the boundary of a particular quadrant. Counting these peaks, the hearer can determine the quadrant in which the topic must be. Because there are not so many objects in the environment, the hearer can determine the topic with reasonable reliability. When the speaker does not rotate, but it does emit IR, the hearer is the topic. When the speaker does not emit IR nor does it rotate, then the speaker is the topic.

The figure, however, represents a graph where the speaker was exactly facing the hearer before it started to rotate. The experiments showed that this method was very unreliable (Steels and Vogt, 1997). The hearer often observed a very different graph than shown and was therefore not able to identify the same topic the speaker pointed at. In the experiments discussed here, this pointing was simulated by means of radio communication. The speaker transferred the quadrant in which the topic was observed to the hearer by radio link. The hearer then maps this quadrant to its own topic. For this mapping, the hearer must first mirror the received quadrant to a quadrant from its own point of view.

When both agents identified the topic and related features to sensory channels for all objects in their surroundings, they can play a discrimination game. The yielded set of distinctive feature sets is then used for the language formation (as described in (Steels and Vogt, 1997)).

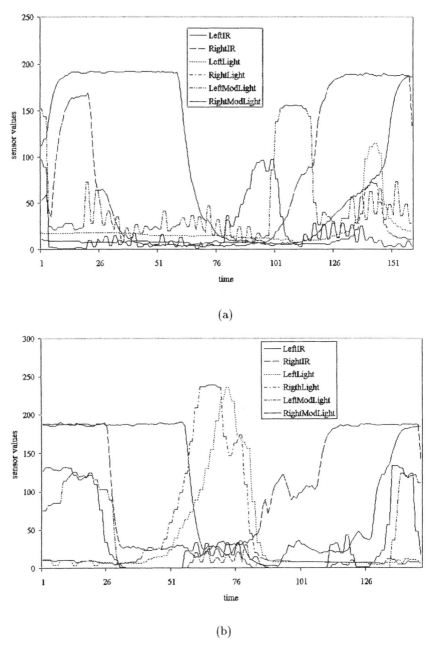

(a)

(b)

Fig. 4. The graph that two opposing robots may sense during the perception. The robots infer that there is an object when two complementary sensors intersect the graph.

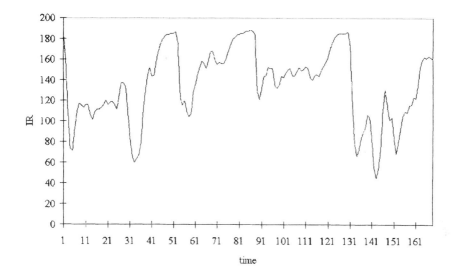

Fig. 5. The graph that the hearer observes when the speaker is rotating 360 degrees, while it is emitting IR in four perpendicular directions.

The speaker and hearer modes can be specified using Finite State Automata where the agents use the radio-link to synchronize the states in which the automata should be, if necessary. Hereto they send messages for asking to communicate, to say that they have aligned etc. The next section reports on the results of the meaning creation and the main features that were important for the successful implementation of the mechanisms.

4 Experimental results

We have done several experiments on the meaning and language formation. This section reports on only one of these experiments. The experiment resulted in 145 naming games, which were the result of two days experimenting. This low amount is due to the following reasons: (1) The robots can only work for approximately 30 minutes when the batteries are charged, and they can complete the process of one naming game roughly every 2 minutes. (2) The internal states (or memory) of the robots and the discourse of the naming games in both robots were monitored by a serial cable. This was extremely difficult since we have two robots, each connected to a cable, that move in an environment full of obstacles. And (3) producing much more games appeared not to be possible because after more games the internal memory started to exceed the 256 kB RAM. This, of course, is the reason why there are not more than 145 games.

As we shall see, the grounding of distinctive features was successful in that the robots revealed distinctive feature sets, which represented the objects rather coherently. This already happened after a short while of learning. The language formation was less clear, because the formation of a language takes much more time. Reports on the language formation can be found in (Steels and Vogt, 1997). This section focuses on the meaning creation.

As was mentioned before, the construction of sensory channels was very important in the development of this experiment. The right choice of these channels must enable the agents to select the right features in order to discriminate the objects coherently. I.e. the selected features, which compose the distinctive feature set(s), must be general enough to be selected in other situations as well. The task for the robots was to ground three objects[1], which are made visible by means of white light, light modulated at a certain frequency and IR. Therefore three sensory channels were constructed for discriminating the objects: *sc0* - *white light*, *sc1* - *modulated light*, and *sc2* - *IR*. All these channels produce values as if the object is seen in the front.

Consider the first discrimination game, where the robot has no features yet. The discrimination game was executed by robot r2, which scanned the object o0 at time/position 67 and object o1 at position 102. The sensory channels for o0 revealed values 69 for sc0, 54 for sc1 and 37 for sc2, for o1 this was 13, 2 and 189 respectively. Obviously, the discrimination ended in failure because there were no features. A new feature is created for sc0, which expects a positive value between 0 and 255.

Discrimination game by r2
Objects r2:
    o0 [67] [69, 54, 37]
    o1 [102] [13,2,189]
Topic r2: o0
Failure r2. No feature sets.
New features r2: r2-sc0 [0,255]

Let us now look at another early discrimination game that fails, but where a feature is divided in two equal halves creating two new features.

Discrimination game by r2
Objects r2:
    o0 [64] [47, 1, 9]
    o1 [103] [76, 46, 182]
    o2 [105] [90, 73, 185]
Topic r2: o0. Feature sets:
    o0 {r2-sc0,r2-sc0-0,r2-sc1,r2-sc2}
    o1 {r2-sc0,r2-sc0-0,r2-sc1,r2-sc2}
    o2 {r2-sc0,r2-sc0-0,r2-sc1,r2-sc2}
Failure r2. No distinctive feature sets.
New features r2: r2-sc1-0 [0,127.5] r2-sc1-1 [127.5,255]

---

[1] The agents need not ground the feature *self*, because this is not an observable feature. It is assumed that this feature is true in every context at time/position 0.

Now we see a discrimination game that was successful. Note that the context is the same as in the previous discrimination game. The robots play several language- and discrimination with the same context (with a maximum of 10).

Discrimination game by r2
Objects r2:
    o0 [64] [47, 1, 9]
    o1 [103] [76, 46, 182]
    o2 [105] [90, 73, 185]
Topic r2: o2. Feature sets:
    o0 {r2-sc0,r2-sc0-0,r2-sc0-0-0,r2-sc1,r2-sc1-0,
        r2-sc1-0-0,r2-sc2,r2-sc2-0}
    o1 {r2-sc0,r2-sc0-0,r2-sc0-0-1,r2-sc1,r2-sc1-0,
        r2-sc1-0-0,r2-sc2,r2-sc2-1}
    o2 {r2-sc0,r2-sc0-0,r2-sc0-0-1,r2-sc1,r2-sc1-0,
        r2-sc1-0-1,r2-sc2,r2-sc2-1}
Distinctive feature sets r2:
    {{r2-sc0,r2-sc1-0-1,r2-sc2},
     {r2-sc-0,r2-sc1-0-1,r2-sc2-1},
     {r2-sc0,r2-sc1-0-1},
     {r2-sc1-0-1,r2-sc2},
     {r2-sc0-0,r2-sc1-0-1},
     {sc1-0-1,r2-sc2-1},
     {r2-sc0-0-1,r2-sc1-0-1},
     {r2-sc1-0-1}}
Success r2. {r2-sc1-0-1} [63.75,127.5]

The 10th discrimination game already showed the above result. Robot r2 finds all combinations of features that include *sc1-0-1*. This feature is the only one that distinguishes o2 from o0 and o1. The game yields $\{r2\text{-}sc1\text{-}0\text{-}1\}$ on the range *[63.75,127.5]*, because this set is the smallest one, which is the preferred condition when there are several distinctive feature sets. The use of all features is increased, whereas the success of only $\{r2\text{-}sc1\text{-}0\text{-}1\}$ is incremented.

This last game illustrates an almost typical view of what the robot observes when it is scanning its surroundings. All sensory channels always yield a noise value, and it is very difficult to say what object in the environment is meant. In order to detect what object is meant, a monitoring program keeps track of what sensory channel is used to decide when the robot observes an object during the perception. In this case the object was recorded because sc2 showed an intersection, therefore o2 is the other robot.

As can be seen in figure 6, discriminative success of the successful distinctive feature sets increase quite rapidly. This is because in the first few discrimination games these sets were used, they were immediately successful. In later games some of these distinctive feature sets were used in a discrimination game, but these were not preferred, thus decreasing the success rate. Some feature sets

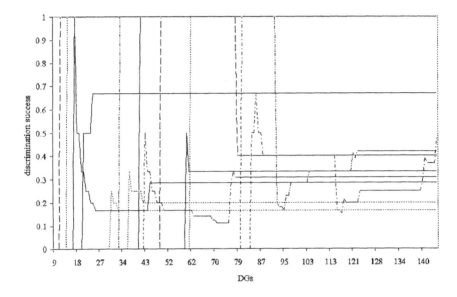

Fig. 6. The discriminative success, i.e. *success/use* of some successful feature sets for the first 145 discrimination games. In the beginning all feature sets are built up with increasing success scores. Some of the score decrease in later games, which is due to competition between distinctive feature sets.

reach a score of 1 immediately. Most of these sets are used only one time after 145 discrimination games.

If we look for which objects the distinctive feature sets are used, we can see that every object has a rather coherent set of distinctive feature sets (table 1). The white light object is recognized with only 16.7% singular distinctive feature sets, but the modulated light objects with 75.0% and the infrared with 72.0%. These figures are, however, taken from only 145 discrimination games, and may therefore not be very significant. Although there is some overlap, most of these distinctive feature sets reveal clear preference for a particular object. To have some overlap and no full coherence in the meaning is likely. It is thought that meaning in the mind is represented by fuzzy sets that reveal a family resemblance (Aitchison, 1994), so that underlying features may reveal some overlap in representing particular objects. Although the experiments have not yet revealed this, it is also thought that the co-evolution of language with meaning will cause all feature sets that now represent one real world object to be associated with one *global* linguistic meaning. So there is normally only one word representing one object, conform (Steels, 1997).

Concluding, we saw that robotic agents were able to learn to discriminate objects using sensory features. The agents build up their own tree of features, and are able to use distinctions in order to recognize the objects. A mechanism of selection for the best used features causes the robot to learn the feature-tree

White light use: 12	Modulated light use: 20	Infrared use: 50
{sc0-0-0}:2	{sc1-0-1}:1	{sc0-0-2}:1
{sc1-0-0}:1	{sc1-0-2}:1	{sc0-1-0}:2
{sc2-1,sc1-0-1}:1	{sc2-0-1}:2	{sc0-0-3,sc1-0-4}:1
{sc2-0-0}:1	{sc2-0-0}:3	{sc1-0-2}:5
{sc2-1-0}:4	{sc1-0-3}:1	{sc1-1-0}:2
	{sc2-0-3}:2	{sc0-0-4}:1
	{sc2-0-4}:2	{sc0-0-5}:9
	{sc2-1,sc1-0-2}:1	{sc2-1,sc1-0-0}:2
	{sc1-0-4}:1	{sc2}:16
	{sc0-0-1,sc2-1-1}:1	{sc2-1,sc0-0-6}:1
	{sc1-0-5}:1	{sc2-1-0}:3
	{sc2-1-1}:3	{sc0-0-0}:5
		{sc0-0-7,sc2-1-1}:1

Table 1. The coherence of distinctive feature sets vs. the objects. The number behind the distinctive feature sets indicates the successfulness of the set for the particular object. Note that the notation of the sensory channels is different than before in order to save some space. Feature $sc1$-0-0 means now the lower half of sensory channel 1, with identifier 0, $sc2$-1-1 is sensitive on the upper half of sensory channel 2, with identifier 1, etc.

in a self-organizing way. Although it has not been shown here, it should be clear that each agent constructs its own set of features, and therefore the agents may have different representations of these features in their 'mind'.

5 Conclusions and future research

This paper discussed the implementation and the results of experiments that has been done on learning to ground semantic features in robotic agents. The experiments were held in combination with language formation experiments, which are reported in (Steels and Vogt, 1997).

We saw that robots were able to construct a feature tree in order to discriminate and categorize sensory information from objects into symbolic feature sets. This categorization was made possible by implementing a self-organizing learning mechanism to create and select features adaptively according to the rules given in section 2. The robots could learn which features were useful to discriminate between objects, thus building coherent classes of feature sets in order to represent these (aspects of) objects, so a lexicon could be formed using these items as semantic features. The feature sets have been grounded completely in distinctions. Real identification of objects (or concepts) though would be another step, which according to Harnad is a necessary, but extremely difficult step in the grounding problem (Harnad, 1990).

Although the results are promising, a lot of research still has to be done. The sets of discriminative features that are given in table 1 must be classified higher up in the hierarchy of meaning. This can be done by the language formation, where agents construct word-meaning pairs that represent discriminative feature sets. It is shown (Steels and Vogt, 1997), that the agents construct ambiguous words (i.e. words with more than one meaning) in the sense that they represent different feature sets. This ambiguity though, is not necessarily visible in the physical world outside an agent, because the different features may represent the same object (see table 1). Therefore we could call this kind of ambiguity a *representational ambiguity*, as opposed to lexical ambiguity and could overcome the identification problem. This principle representational ambiguity, however, has not revealed in the experiments sufficiently, because the formation of a coherent lexicon may take more than a thousand language games. Due to technical problems experiments at such a scale were not yet executed.

Currently we are working on the improvement of these robotic experiments, so we will be able to execute these large-scale experiments. We are also preparing experiments to ground and lexicalize spatial relations, internal states and actions. Work is being carried out in order to implement grounded language formation using active vision.

Concluding, we could say that this experiment might be an important step towards a new theory of meaning and language, where the formation is based on a natural selection-like principle. The basic building blocks of these languages are not innate structures (conform (Chomsky, 1980)), but rather the result of (1) life-long evolution inside an agent based on its interactions with its environment (Foley, 1997) and (2) the social interaction between different agents in a group.

6 Acknowledgements

This experiment was done as a project for my master's thesis in Cognitive Science and Engineering at the University of Groningen. It was carried out as a practical period at the VUB Artificial Intelligence Laboratory in Brussels.

I would like to thank my supervisors Luc Steels (VUB) and Petra Hendriks (Univ. of Groningen) for their support and advice during the project. Furthermore I would like to thank the members of the AI-Lab for their support: Tony Belpaeme, Andreas Birk, Bart de Boer, Edwin de Jong Peter Stuer and Dany Vereertbrugghen for advice and support.

The project was partly financed by an EU Erasmus fellowship.

References

Aitchison, J. (1994). *Words in the mind: an introduction to the mental lexicon.* Blackwell Publishers, Cambridge Ma., second edition.

Birk, A. (1997). Autonomous recharging of mobile robots. In *Proceedings of ISATA97, Dedicated Conference on Robotics, Motion and Machine Vision.*

Braitenberg, V. (1984). *Vehicles, Experiments in Synthetic Psychology.* The MIT Press, Cambridge MA.

Chomsky, N. (1980). Rules and representations. *The behavioral and brain sciences*, 3, pp. 1-61.

Edelman, G. (1987). *Neural Darwinism.* Basic Books Inc., New York.

Foley, W. (1997). *Anthropological linguistics : an introduction.* Blackwell, Cambridge Ma.

Harnad, S. (1990). The symbol grounding problem. *Physica D*, 42, pp. 335-346.

Prigogine, I. and Strengers, I. (1984). *Order out of Chaos.* Bantam Books, New York.

Steels, L. (1994). Artificial life route to artificial intelligence. *Arificial Life Journal*, 1(1).

Steels, L. (1996a). Emergent adaptive lexicons. In P., M., editor, *From Animals to Animats 4: Proceedings of the Fourth International Conference On Simulating Adaptive Behavior*, Cambridge Ma. The MIT Press.

Steels, L. (1996b). Perceptually grounded meaning creation. In M., T., editor, *Proceedings of the International Conference on Multi-Agent Systems*, Menlo Park Ca. AAAI Press.

Steels, L. (1996c). The spontaneous self-organization of an adaptive language. In Muggleton, S., editor, *Machine Intelligence 15*, Oxford. Oxford University Press.

Steels, L. (1997). Synthesising the origins of language and meaning using co-evolution, self-organisation and level formation. In Hurford J., C. K. and Studdert-Kennedy, M., editors, *Evolution of Human Language*, Edingburgh. Edinburgh University Press.

Steels, L. and Vogt, P. (1997). Grounding adaptive language games in robotic agents. In Husbands, C. and Harvey, I., editors, *Proceedings of the Fourth European Conference on Artificial Life*, Cambridge Ma. and London. MIT Press.

Vereertbrugghen, D. (1996). Design and implementation of a second generation sensor-motor control unit for mobile robots. Master's thesis, VUB.

A Learning Mobile Robot:
Theory, Simulation and Practice

Nuno Chalmique Chagas and John Hallam

University of Edinburgh, Department of Artificial Intelligence,
5 Forrest Hill, Edinburgh EH1 2QL, UK
{nuno,john}@dai.ed.ac.uk

Abstract. We describe the implementation and testing of the KINS multi-strategy learning controller for real mobile robot navigation. This controller uses low-level reactive control that is modulated on-line by a learning system based on case-based reasoning and reinforcement learning. The case-based reasoning part captures regularities in the environment. The reinforcement learning part gradually improves the acquired knowledge. Evaluation of the controller is presented in a real and in a realistic simulated mobile robot, across different types of environments.

1 Introduction

How to specify behaviour in a robot has come a long way since the low-level languages of assembly robotics [Lozano-Perez, 1982]. The *classical AI* approach to control[1] proved too slow and too fragile for the real world but showed that representations of the environment, however difficult to maintain, produce interesting behaviour. In *nouvelle AI, e.g.*[Brooks, 1985, Brooks, 1991a, Brooks, 1991b], agents merely react to the current environmental situation posed, limited to the scope of the designer's built-in behaviours which are very hard-wired. However, low-level behaviours are very good at being fast, like reflexes.

Learning in mobile robot controllers is advantageous because it lessens the burden of exhaustive specificity. Learning can help make continually updated representations of the environment, as well as optimise low-level behaviours. Robot controllers that learn are more flexible and more prone to succeed at whatever task they are meant to do.

The solution that is the object of this paper is the Khepera Improving Navigation System — KINS. This is a practical learning controller, applied to a real mobile robot.

KINS is the product of extensive modification to the Self Improving Navigation System — SINS — [Ram and Santamaria, 1993b, Moorman and Ram, 1992] which was done in a simple computer simulation only. SINS is a highly competent controller which has proven in simulation to be a good hybrid solution with good results incorporating advantages from the different techniques it uses:

---

[1] See [Pfeifer and Scheier, 1994] for a discussion.

reactive control, reinforcement learning and case-based reasoning. It is also capable of temporal learning because it encodes acquired knowledge as structured sequences through time.

KINS is a case-based learning system on top of a reactive system that carries on from SINS as a blend of reactive/learning strategies. What appeals most is that the designer knowledge that is actually put into the system — the reactive behaviours — is plastic and controlled on-line by a learning scheme allowing quite complex behaviour to be learned with no bias from the designer and relatively low effort to program.

In [Brooks and Mataric, 1993], robot learning is divided into four types. KINS achieves the first three of these:

- *Learning of Numerical Functions for Calibration or Parameter Adjustment.*
  Operational parameters of an existing behavioural structure are optimised during the life of the robot.
- *Learning about the World.*
  Internal functional representations of the world are constructed and altered.
- *Learning to Coordinate Behaviours.*
  Existing behavioural structures and their effects in the world are used to change the conditions and sequences in which they are triggered under control of internal learned functions.
- *Learning New Behaviours.*
  KINS does not learn new behavioural structures, as such. However, the behaviours are configurable by parameters, which adds a degree of plasticity.

The next section is a description of the KINS architecture. Section 3 describes the testing methodology used, its purpose and the results obtained. Finally section 4 ponders on the results obtained.

2 The KINS architecture

The KINS — Khepera Improving Navigation System — is an unsupervised learning controller architecture that makes use of reactive control, case-based reasoning and reinforcement learning.

It is composed of two modules (figure 1a):

- the Reactive Control Module
- the On-line Adaptation and Learning Module

Both modules have free access to the sensory information of the robot and run concurrently. The first module is a collection of parameterisable reactive behaviours which account for the immediate responsiveness of the system. The second one is a case-based and reinforcement learning system that learns sets of numerical functions of time — cases (outlined in Figure 1b) — that output parameter values for the low-level behaviours of the first module.

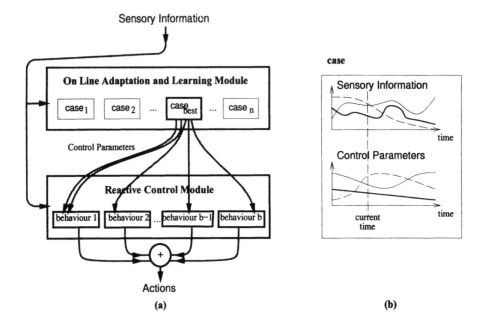

Fig. 1. KINS Architecture

Each function in a case represents either a sensory information variable or a control parameter. Each control parameter function in a case controls one parameter of a behaviour. Each sensory information function in a case represents a history of sensory *indicator* values. An indicator is a function of the sensors of the robot, *e.g.* an obstacle-density indicator can reflect if the robot has obstacles around it, based on proximity sensor information. Indicators are predefined by the designer and fixed. Thus, a case is a series of values in time which associate behavioural regulatory parameters to sensory information, aiming for the most satisfactory navigation efficiency in a particular contextual situation. The learning of cases is made from scratch, *i.e.*there are no cases to start with, and these are gradually created, improved on and extended in time.

In the reactive control module, all of the behaviours run in parallel. They are given independent access to the vehicle's sensors and compute a command vector[2]. All the vectors, one per behaviour, are simply summed up to produce the final vector which commands the robot. Each behaviour has at least one parameter that regulates its operation — that is the magnitude or gain of its output vector. Each behaviour may have other parameters that can be modulated, pertaining to its function. All of the behaviours' parameters are dynamically set by the on-line adaptation and learning module, while the robot is in operation.

The goal of KINS is to navigate efficiently to a target position through a

---

[2] consisting of the left and right motor commands

cluttered environment. Efficiency is attained by three components which make up the reinforcement function: moving fast, not bumping and progressing towards the goal.

It has a reactive control module with three behaviours:

- *Move-To-Goal* — produces a vector in the direction of the goal;
- *Avoid-Obstacles* — produces a vector away from obstacles;
- *Noise* — produces a new random vector from time to time, making the robot describe a jagged-line trajectory. It has NOISE-PERSISTENCE as a parameter which controls how often a new random vector is generated.

There are thus four parameters that the on-line learning and adaptation module controls. These are NOISE-PERSISTENCE (just described) and the gains to each of the behaviours — MOVE-TO-GOAL-GAIN, AVOID-OBSTACLES-GAIN and NOISE-GAIN. The parameters take analogue values in a bounded range. The set of these four parameters determines at each moment the composite behaviour the robot actually assumes, by allowing each behaviour's influence on the total behaviour of the robot to take any value from a maximal value through to zero.

There are four indicators in KINS:

- *Obstacle-density* reflects how much free space there is in the immediate vicinity of the robot;
- *Motion-to-goal* reflects whether the robot is progressing towards or away from the goal;
- *Relative-motion* reflects how much the robot is moving, any motor activity;
- *Absolute-motion* reflects how efficient the motion of the robot is, *i.e.*is low if the robot is moving back-and-forth.

Thus, each case consists of eight functions in time, all of the same length. Four of these sequences are indicators. Four of them are behavioural parameters (figure 1b).

The On-line Learning and Adaptation Module consists of a loop of the following four phases:

- *Perceive* — Monitors the sensors of vehicle and computes current indicator values;
- *Retrieve* — Detects and discriminates between different environmental configurations by searching exhaustively the collection of learned cases for the case that best matches the robot's own history of indicator and control parameters — contextual environmental situation history. A *goodness of match* measure is computed for every learned case to reflect how similar this, or part of this case, is to the contextual environmental situation history, and at which point in the case timeline;
- *Adapt* — Sets the behaviours' parameters in a dynamic on-line manner based on the control parameters for the succeeding point to the best matching point in time of the best matching case. This is based on the assumption that a similar situation is being encountered and that the case has led to positive results in the past;

– *Learn* — Simultaneously updates the internal representation of the case according to current experience. A more detailed description of this phase follows.

The learning phase involves three different methods of incorporating new knowledge. How relevant a particular learned case is to the particular environmental situation is determined directly by the goodness of match measure — the better the match, the higher the *relevance*. According to relevance a case can either be:

– *Created* — Creation of a new case based on the current situation is performed if the relevance of all existing cases is minimal;
– *Modified* — Modification of the associations made in a case, if relevance is acceptable. A case is modified either by being reinforced or explored depending respectively on how well or how badly the robot is performing. A case is reinforced by making the associations in the case more like the robot's contextual environmental situation history. A case is explored by altering the associations in the case by a random factor. Both types of case modifications — reinforcement and exploration — are done proportionally to the evaluation of the reinforcement function. This allows several degrees of reinforcement or exploration. They are also done back through the time line of the case, i.e., they are done as a weighted regression back in time, where the closer to the present, the more influential a case modification is.
– *Extended* — Extension of the size of a case to yield associations over larger time windows is performed if relevance is very high, by extending the last association in the case a further step.

2.1 KINS and reality

KINS as described so far includes a number of important differences from SINS from which it derived. These modifications were primarily driven by the need to run on a real robot — namely the Khepera mobile robot and its simulator [Michel, 1996] — instead of on a Cartesian computer simulation. These changes were imperative when the architecture required global world information. Such is the case of Cartesian movement, relative motion and above all absolute motion.

A list of the changes made followed by a brief description of each is given below.

– *The relative and absolute motion indicators* are based on dead reckoning from left and right encoders, and not from global X and Y coordinates. This may produce an effect that never happened in Cartesian simulation, since the X and Y coordinates of the robot were not prone to error as dead-reckoning is.
– *The behaviours' output* are left and right motor commands and not Cartesian vectors. This proved to be particularly awkward with the random behaviour which has to repetitively turn on the spot for a time and then go in a straight line for some more time. In the Cartesian simulation the robot did not need to turn.

- *The goal* is represented by a light. The distance and bearing to the goal can only be measured by the robot's light sensors which have limited accuracy and range.

- *The Move-To-Goal behaviour* is somewhat different from the original, in that a constant gradient of light is not always present, so, often, the robot doesn't know which way to go. In this case, the behaviour now tells it to go forward, even though it may be getting further away from the light source.

- *The bump sensor* was programmed to be triggered if any of the distance sensors saturates. This is necessary because there is no physical bump sensor on the Khepera.

- *The reinforcement function* was changed from a function that penalised bumps and low absolute-motion, to a function that also rewards motion-to-goal.

- *The reinforcement procedure* now depends on the reinforcement function magnitude, rather than just sign, introducing a notion of analogue values of case reinforcement and exploration.

3   Experiments

Three sets of experiments were made, each addressing a specific problem:

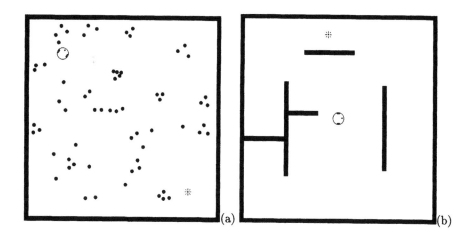

Fig. 2. (a) Cluttered environment; (b) Maze environment

1. KINS *vs.* SINS — by experimenting with the Khepera simulator in a cluttered environment and comparing with the results in [Ram and Santamaria, 1993b];

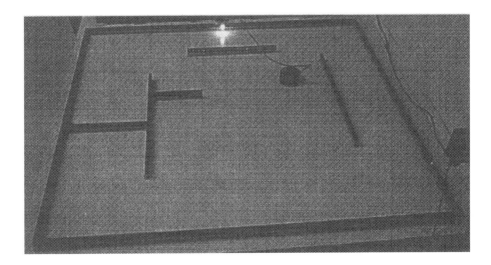

Fig. 3. Practical Setup - LEGO maze environment, Khepera robot and light (goal)

2. *Effect of Maze environment on navigation* — as opposed to cluttered environment navigation (see figure 2) — the maze environment is fundamentally different from the cluttered environment and, tentatively, is more likely to be encountered by a real mobile robot;
3. *Effect of running on a real robot* — by comparing the Khepera simulator results for the maze environment (figure 2b), with the real Khepera performance in an equivalent, real maze environment (figure 3).

The performance evaluation is based on the concept of *run*. A run is a trajectory from a new random starting point to the light (goal). A limit of 4000 steps per run was set which, if reached, means that the run was not solved. The optimal distance is the distance measured in a straight line from the starting point of the run to the goal.

For the comparison of results, five estimators were used to evaluate the performance:

1. *Percentage of runs solved*
2. *Average steps per run*
3. *Average distance travelled per run*
4. *Average $\frac{travelled}{optimal}$ distance per run*
5. *Average collisions per run*

These estimators were also used in [Ram and Santamaria, 1993b] and provide a direct way of comparing results.

Three types of controllers were used to test the efficiency of the on-line learning and adaptation module:

- KINS — the case-based reasoning and reinforcement learning on-line learning and adaptation module;
- CONSTANT — which yields constant behaviour parameters. All of the behaviours' parameters, namely MOVE-TO-GOAL-GAIN, NOISE-GAIN, NOISE-PERSISTENCE and AVOID-OBSTACLES-GAIN are set to $\frac{1}{2}$. The range for these parameters is $[0, 1]$.
- RANDOM — which continually sets behavioural parameters randomly.

Each controller was tested for 200 runs with later runs benefiting from earlier experience. This procedure was performed 20 times, starting with reset controllers and the mean performance of each set of corresponding runs was obtained. The functions in the graphs are running means from start to current run as in [Ram and Santamaria, 1993b].

Controller	% runs solved	steps per run	distance per run	$\frac{travelled}{optimal}$ distance per run	collisions per run
KINS	72.25%	2395	4377	7.92	257.8
SINS*	100.00%	141.9	107.2	2.63	10.7
ratio	**0.72**	**16.9**	**40.8**	**3.0**	**24.1**
CONSTANT	38.28%	3096	3655	6.57	957.5
STATICH*	82%	315.5	126.2	3.10	48.1
ratio	**0.47**	**9.81**	**29.0**	**2.1**	**19.9**
RANDOM	72.17%	2268	4710	8.46	266.1
RANDOM*	100.00%	141.0	113.7	2.79	13.7
ratio	**0.72**	**16.1**	**41.4**	**3.0**	**19.4**

Table 1. Performance in the cluttered environment with a "de-fogged" light for the simulated robot. *results from [Ram and Santamaria, 1993b].

3.1 KINS vs. SINS

The experiments reported in [Ram and Santamaria, 1993a] were done in the Khepera simulator. A cluttered world (figure 2a) was used and the light was "de-fogged" so that it could be seen from a long distance. This allows a gradient of light to be felt from any part of the arena as in SINS the goal could always be seen. This was made so that the comparative study is more faithful[3].

---

[3] Despite every effort to make the testing conditions similar, a significant difference

150

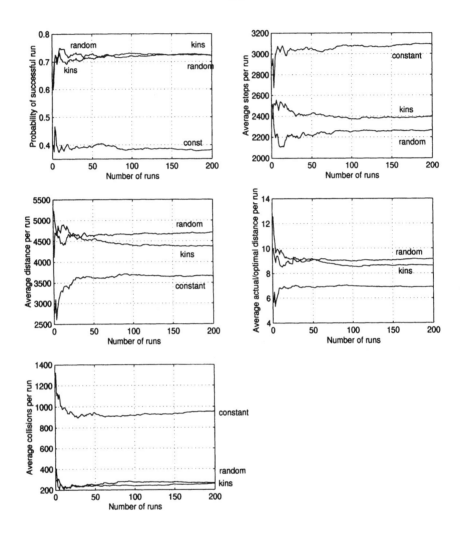

Fig. 4. Khepera simulator results for the cluttered environment with "de-fogged" light.

Table 1 shows the performance of the three different types of controllers, each for the Khepera simulator and for the Cartesian simulator. As the two implementations of estimators differ, a ratio is presented for each pair of comparable performance estimators.

exists between the testing methodologies. While the KINS robot is started from different starting points for each run and the world is always the same, the SINS robot's starting and goal positions remained the same throughout the runs and a new world was generated randomly after each run. This makes the KINS results more variable from run to run.

From the ratios it can be seen that the Khepera simulator implementation of the KINS and RANDOM controllers, conformed to the expectations, agreeing to a comfortable extent with the homologous controllers' results obtained in the Cartesian simulator. The CONSTANT controller behaves quite differently from the STATICH constant heuristic controller.

Figure 4 shows the evolution of the results through the runs for the Khepera simulator implementations of the three controllers in the cluttered environment. It can be concluded that KINS performs best, closely followed by RANDOM and that CONSTANT performs badly solving only under 40% of the runs and bumping 4 times as much as KINS.

3.2 Effect of maze environment on navigation

After this, and with the purpose of having simulated results that could be more faithfully compared with the real-robot results in the next subsection, the light was made more "real" in that it could only begin to be seen in a local vicinity which is circa $\frac{1}{9}$ of the area of the environment. The maze environment shown in figure 2b was used. This particular maze configuration creates a local minimum when the robot is at the location near the light with the wall between the two, because the robot can see the light through the wall. This is a special pitfall for any controller since it can only escape from that particular zone by means of a surge of the random behaviour.

Controller	% runs solved	steps per run	distance per run	$\frac{travelled}{optimal}$ distance per run	collisions per run
KINS	98.57%	1017	2402	4.69	5.9
CONSTANT	98.70%	916	2332	4.50	29.6
RANDOM	95.73%	1207	2864	5.62	16.0

Table 2. Performance in maze environment using simulated robot.

The results in this simulated maze environment are shown in table 2 and figure 5. KINS converges to be as successful as CONSTANT, which are both considerably better than RANDOM. The fundamental difference between KINS and CONSTANT is that KINS bumps only $\frac{1}{5}$ of the times CONSTANT does, clearly giving KINS the edge over CONSTANT.

KINS converges fast after about 50 runs (figure 5). After this it manages to hold its performance. KINS learns very efficiently not to collide surpassing noticeably all other controllers on this indicator, bumping only 20% of the times CONSTANT does and 50% of the times of its nearest competitor (RANDOM). This is at some expense of the steps per run measure, which is only 10% higher than

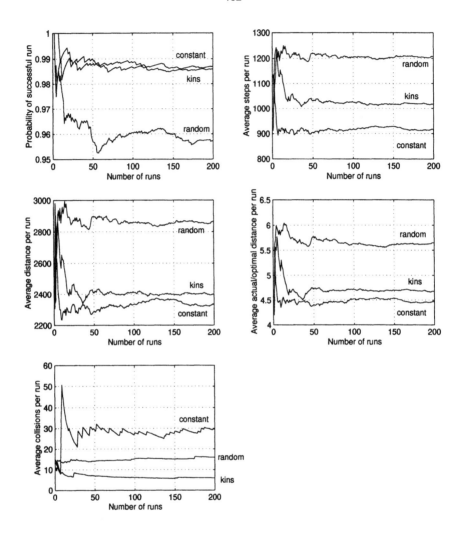

Fig. 5. Khepera simulator results for the maze environment.

the CONSTANT controllers' and still 20% lower than RANDOM's. KINS learns to navigate at lower velocities than CONSTANT and so, takes more time steps to reach the goal. This is because it learns to care about obstacles and navigates at sensible speeds which allow it to minimise collisions. On a trajectory analysis, KINS gets into tighter spots than the other controllers, and manages to come out of them, proving to be a more thorough explorer.

153

3.3 Effect of running on a real robot

From the Khepera simulator, experiments were then made with a real Khepera in a LEGO environment. The practical setup can be seen in figure 3. It has the same configuration as the simulated environment in figure 2b. The walls have small round holes which let the light through. This is in accordance with the simulator too, in which obstacles and walls are "transparent" to light.

For this experiment 100 runs were made with the KINS controller. For the CONSTANT and RANDOM controllers, only 20 runs were made since there is no adaptivity in these controllers. There were 20 different starting points. The KINS controller used each one of these starting points 5 times, while the other controllers used each starting point only once.

The practical results[4] attained are shown in table 3. They can be compared with those of table 2. The results indicate that KINS holds up in the Khepera robot and manages to outperform the CONSTANT and RANDOM controllers in the average of the 100 runs, and especially in the last 10 runs.

The CONSTANT controller failed to reach the light once because it was unable to negotiate the wall by the light, getting stuck. This major drawback was anticipated from the Khepera simulator results, due to the enormous propensity of the CONSTANT controller to bump.

Also, it is is worth mentioning that the disadvantage KINS suffers in computing time against the RANDOM and CONSTANT controllers goes by unnoticed in the real robot implementation because of the time the robot needs to physically move.

Controller	solved runs/ total runs	steps per run	distance per run	collisions per run
KINS	100/100	1039	2523	11
KINS$_{10}^{**}$	10/10	764	1895	9
CONSTANT	19/20	1040	2529	16
RANDOM	20/20	1207	2694	13

Table 3. Performance in maze environment using real robot. **Last 10 runs of KINS (runs 91 to 100).

4 Conclusion

A model of a controller capable of efficient adaptive navigation for a mobile robot — KINS — was presented and its efficiency evaluated.

The controller was based on SINS and was extensively modified in order to work on a real robot simulator and on a real robot. The improbable toy worlds of simulation and SINS were replaced by a real Khepera world and KINS.

---

[4] The optimal distance for each run was not recorded because it cannot easily be measured in practice so the average $\frac{travelled}{optimal}$ distance per run estimator is not presented.

Considerations that extend to the simulations done are that KINS managed to better non-learning controllers in a cluttered and in a maze environment.

Experimental results obtained with a real mobile robot in a maze environment confirmed that the average performance of KINS, starting with no knowledge, bettered that of similar but non-learning controllers in 100 runs. In particular, after having learned, KINS clearly outperforms both non-learning controllers.

Acknowledgements

Nuno Chagas is supported by PRAXIS XXI grant BD/2948/94. Thanks to an anonymous reviewer and to Sandra Gadanho for their helpful comments. Facilities were provided by the Department of Artificial Intelligence of the University of Edinburgh.

References

[Brooks and Mataric, 1993] Brooks, R. and Mataric, M. (1993). Real robots, real learning problems. In Connell, J. and Mahadevan, S., editors, *Robot Learning*, chapter 8, pages 193–213. Kluwer Academic Publishers.

[Brooks, 1985] Brooks, R. A. (1985). A robust layered control system for a mobile robot. AI Memo 864, MIT.

[Brooks, 1991a] Brooks, R. A. (1991a). Intelligence without reason. AI Memo 1293, MIT.

[Brooks, 1991b] Brooks, R. A. (1991b). Intelligence without representation. *Artificial Intelligence*, pages 139–159.

[Lozano-Perez, 1982] Lozano-Perez, T. (1982). Robot programming. AI Memo 698, Massachusetts Institute of Technology - Artificial Intelligence Laboratory.

[Michel, 1996] Michel, O. (1996). Khepera simulator package version 2.0: Freeware mobile robot simulator. Technical report.

[Moorman and Ram, 1992] Moorman, K. and Ram, A. (1992). A case-based approach to reactive control for autonomous robots. In *AAAI Fall Symposium on "AI for Real-World Autonomous Mobile Robots"*, Cambridge, MA.

[Pfeifer and Scheier, 1994] Pfeifer, R. and Scheier, C. (1994). From perception to action: The right direction? In Gaussier, P. and Nicoud, J.-D., editors, *From Perception to Action Conference*, pages 1–11. IEEE Computer Society Press.

[Ram and Santamaria, 1993a] Ram, A. and Santamaria, J. C. (1993a). A multistrategy case-based and reinforcement learning approach to self-improving reactive control systems for autonomous robotic navigation. In *Proceedings of the Second International Workshop on Multistrategy Learning*, WV. Harpers Ferry.

[Ram and Santamaria, 1993b] Ram, A. and Santamaria, J. C. (1993b). Multistrategy learning in reactive control systems for autonomous robotic navigation. *Informatica*, 4(17):347–369.

Learning Complex Robot Behaviours by Evolutionary Computing with Task Decomposition

Wei-Po Lee, John Hallam, Henrik Hautop Lund

Department of Artificial Intelligence,
University of Edinburgh,
Edinburgh EH1 2QL, Scotland, UK
email:{ weipol,john,henrikl} @dai.ed.ac.uk

Abstract. Building robots can be a tough job because the designer has to predict the interactions between the robot and the environment as well as to deal with them. One solution to cope the difficulties in designing robots is to adopt *learning* methods. Evolution-based approaches are a special kind of machine learning method and during the last few years some researchers have shown the advantages of using this kind of approach to automate the design of robots. However, the tasks achieved so far are fairly simple. In this work, we analyse the difficulties of applying evolutionary approaches to learn complex behaviours for mobile robots. And, instead of evolving the controller as a whole, we propose to take the control architecture of a behavior-based system and to learn the separate behaviours and the arbitration by the use of an evolutionary approach. By using the technique of task decomposition, the job of defining fitness functions becomes more straightforward and the tasks become easier to achieve. To assess the performance of the developed approach, we have evolved a control system to achieve an application task of box-pushing as an example. Experimental results show the promise and efficiency of the presented approach.

1   Introduction

In recent years, building reactive control systems for robots has become a major alternative to traditional robot design. This approach has been proven to be able to achieve real-time performance for robots by creating tight coupling between perceptions and actions. However, predicting and dealing with unforeseen situations and circumstances in the environment make the design still difficult. Consequently, the idea of getting the robot to learn to achieve the tasks, through the interaction between the robot itself and the environment, is advocated.

Evolutionary Robotics is a kind of approach which enables the robot to learn behaviours by the use of evolutionary techniques. This approach differs from other learning skills in that it operates a population of agents rather than a single one. This kind of approach has recently attracted much attention; a lot of work has been conducted to evolve robot controllers and their preliminary results have shown the promise of this approach (e.g.,[3][6][12]). Yet, the tasks achieved so far, such as obstacle avoidance or light seeking, are relatively simple. To help the human designer to develop control systems which are difficult to handcode, we have to resolve the problem of how to scale up the evolution-based approach in controlling robots. In this work we discuss, from different points of view, some of the difficulties one will encounter in evolving controllers to accomplish complex tasks. We also suggest that task decomposition is an efficient technique in supporting the evolutionary approach to evolve controllers for complex tasks. To demonstrate this, we undertake an application task, in which a robot is required to push a box toward a specific position indicating by a light source. This task is fairly complex compared to the ones achieved in previous Evolutionary Robotics experiments and the results show that the robot can achieve the specified task reliably.

2 Evolving Controllers to Achieve Complex Tasks

2.1 The Difficulties

Generally speaking, the evolutionary approach is a kind of search-based approach in which genetic operators, such as reproduction, crossover, and mutation, are used and expected to find a satisfactory solution from a hyper-space; and the dimension of this space is determined by the length of the chromosome. For instance, in a binary encoding scheme, a chromosome with length n indicates that the evolution techniques are expecting to find the appropriate solution from a space with 2^n candidates. Thus, when the length of the chromosome is reasonably increased in respect to the increment of task complexity, the solution space will grow exponentially and lead the search to be more and more difficult. This is particularly apparent in the work that uses recurrent neural networks as control systems since the characteristic of recurrence leads the length of chromosome to increase quadratically and thus enlarges the search space even faster.

The increasing task complexity also introduces difficulty in defining fitness functions to guide the search direction during the evolution. In general, the increment of task complexity implies a higher-level goal to achieve, which almost always involves the interaction of multiple subgoals. For a complex task, directly defining a fitness function at the higher-level is relatively simple but it makes the task difficult to be achieved. On the contrary, defining fitness functions at lower-level is more difficult while it makes the task more achievable. For example, in the work [13], the authors have shown that in their grasping task, if the fitness function was simply defined as the number of objects grasped and deposited correctly, then the desirable behaviour could not be evolved successfully. This is due to the fact that during the earlier generations none of the individuals could achieve the complete task; it resulted in the equally bad fitness for all the populations (all scored zero) and made all the control systems indistinguishable in performance. On the other hand, if lower-level subgoals were introduced to the fitness function, such as rewarding the behaviours of recognizing objects and picking objects up, the performance of controllers became more distinguishable and then the target was achieved. Manipulating fitness at lower levels can assist the evolutionary system to converge; however defining an appropriate fitness function at a lower level is never easy because it has to deal with the multiple subgoals simultaneously. Further, this kind of difficulty occurs in consequence of the increment of the task complexity.

On the other hand, from the point of view of controlling a robot, one may want the evolved control systems to be distributed for their corresponding advantages. In a distributed architecture, the perceptual processing is distributed across multiple independent modules, and every module only deals with the sensory information directly related to its particular need. This not only reduces the sensory bottleneck but also allows each control module to be developed with the most suitable representation and approach with least restriction. Owing to the modular and distributed characteristics, the performance of the overall system will degrade gradually, even if some of the devices or control strategies do not function properly. Further, with an explicitly distributed architecture, an overall system will be easily integrated from different subsystems which could be designed independently; it can also be easily maintained. Therefore, from the point of view of developing systems for robot control, distributed control architectures are preferred.

2.2 Task Decomposition

In order to reduce the search space to make the search easier, to simplify the job of defining fitness functions, and to obtain a distributed control system, a promising way is to adopt the divide-and-conquer problem-solving methodology. In this kind of approach, the designers break tasks from complex (higher-level) down to simple (lower-level) and then achieve the tasks in the reverse sequence. How to decompose a task of course depends on the designers' experiences, but human designers are normally quite skillful in doing it. The tasks are arranged to be achieved in the sequence of increasing complexity and, at each level, the control systems are evolved on top of the ones evolved at lower-levels. Hence, fitness functions will become easier to define and the tasks will be easier to achieve (the fitness function of a certain level task can be defined simply as the goal at this level, to reduce the difficulty in embedding the lower-level subgoals into it; and evolving control systems on top of other lower-level controllers can exploit their corresponding control skills to achieve the current goal). In addition, each subtask only needs to deal with the perceptual information directly related to it, which also makes the tasks easier to achieve. In the robot learning domain, some work has shown that the decomposition technique helps in achieving more complex tasks [4][11].

Actually, the concept of this kind of approach is much like behaviour-based control, which has been successfully and widely used in the robot community for building autonomous robots (e.g.,[2][14][17]); while the main difference is that the approach here employs evolutionary techniques to *evolve* new behaviours and behaviour coordinators rather than to *handcode* them. With evolutionary techniques, the human designer can concentrate on the system level design and let the evolutionary system take care of the implementation details. In addition, since in this approach the tasks are decomposed along the horizontal way as proposed in [2], the corresponding control architectures will be explicitly distributed and can fully exploit all the advantages of distributed architecture as analysed in the section above.

We are particularly interested in investigating ways to reduce the load of robot programmers and in evolving distributed architectures for complex tasks. Because, at the present stage, the task-decomposition technique seems to be the most direct way to achieve these, we will concentrate on investigating how to use this approach, with our genetic programming (GP) system [8][9], to evolve control modules and coordinators to achieve complex tasks.

3 Evolving Hierarchical Task-achieving Controllers

As described above, in order to evolve distributed control systems to achieve complex tasks, we intend to use the technique of task decomposition to break the overall tasks and use the GP techniques to evolve separate behaviour controllers and coordinators for integration. In this section, we will explain the aspect of the control architecture corresponding to the task decomposition, and then describe the genetic representation of the control module to be evolved.

3.1 Control Architecture

Since we will decompose tasks in a hierarchical way, the corresponding control system is organized as multiple layers. After the decomposition, the control system includes a set of *behaviour primitives* and *behaviour arbitrators*. Here, a behaviour primitive is a reactive controller with the representation described in the below section; it involves the lowest level sensory-motor control. Unlike the priority network in the subsumption architecture [2], a behaviour arbitrator here is not hardwired in advance; it is also treated as a reactive controller. The behaviour arbitrator has the same structure and representation as the primitives; the only difference between them is that the output of a primitive is used to control the motors while the output of an arbitrator is used to activate one of the sub-controllers involved. Thus, similar to a reactive planner [5] or a conditional sequencer [7], an arbitrator here allows the binding between environment conditions and activations of lower level behaviours to take place at the run time. This provides adaptiveness not only at the lower level sensory-motor control but also at the behaviour level.

Because our computing system does not support parallel computation, all the control modules are passive and the control flow is thus from top to down. At each time step, the highest level arbitrator evokes one of the related sub-controllers to be in charge the control, according to certain sensory information. If the evoked sub-controller includes an arbitrator, this arbitrator will be evaluated first and its output can then be used to activate another controller. This process continues until a control primitive at the lowest level takes control and drives the actuators. Figure 1 illustrates the general architecture of our control systems and the implementation of an arbitrator.

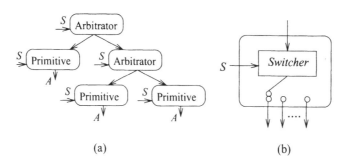

(a)   (b)

Fig. 1. (a) The general architecture of a control system. S and A represent the sensors and actuators related to a certain control work; (b) the implementation of an arbitrator.

3.2 Representation

When using evolutionary computation techniques to solve a problem, the first important step is to choose the proper representation for an individual. On the one hand, a genetic representation must be able to express explicitly the features of the solution of the problem to be solved; on the other hand, it must be suitable to be manipulated by the genetic operators to obtain the solution. The following sections are about how we develop the genetic representation of the reactive controllers to be evolved in this work.

The Circuit Model of a Behaviour Controller A promising choice to satisfy the above requirements is the *circuit network* which has been proven to provide a finer-grained view to represent a behaviour controller. In the *circuit approaches* [1][15][16], an agent (behaviour controller) exists in the form of digital hardware; and it is constructed by two types of components, pure functions and the delays, depending on what kind of tasks (reactive or sequential) it is achieving. Pure functions are logic gates, and delays correspond to flip-flops or registers. The output of one component may be input to one or more other components, thus forming a network. Signals propagate through the network and sensing is thus linked to action. As is well known, any finite state transduction can be carried out by such a network.

Since this work focuses on evolving reactive controllers, we will only discuss how to evolve controllers of this kind. The approach can be extended to evolve sequential ones with minor modification.

The genetic representation of our reactive controller is inspired by the logic representation in the circuit approaches. By duplicating and separating the components of which the outputs serve as inputs of multiple components and by introducing a dummy root node to connect the outputs of a circuit network together, we find it very straightforward to convert a circuit network to a circuit tree. Figure 2 shows an example. After structuring information from the environment and defining simple syntactic rules properly, we can use a GP system to evolve the circuit trees.

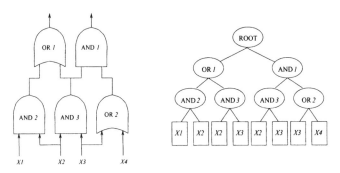

Fig. 2. An example shows converting a circuit network to a tree.

Genetic Representation of Our Reactive Controllers According to our design, the perception information is structured as *sensory conditionals* and connected to the inputs of a logic circuit. The structured sensor-conditionals involve comparing the responses of different sensors or comparing sensor response to numerical thresholds. For these purposes, both sensor responses and numerical thresholds are normalised to be between 0 and 1 inclusive. Thus, a sensor conditional has a constrained syntactic structure; it exists in the form of X >= Y, where X, Y can be any normalised sensor response or threshold which is determined genetically. Figure 3 shows the representation of our reactive controllers.

After organizing our genetic representation, we can classify the involved symbols into *terminals* and *non-terminals* for manipulating a GP system. The dummy root node, the logic components, and the comparator are defined as the non-terminals; while the normalized sensor responses and numerical thresholds, which constitute the sensory conditionals, are defined as terminals. In order to evolve the controllers with the above representation to solve different control tasks, we will need to define different sensor terminals depending on the requirements of the specific task. The experiment sections of this work will give the details of how to evolve such kind of controllers.

4 Experimental Setup

4.1 Application Task

In the following experiments, we will follow the approach described above to evolve control systems for a moderately difficult box-pushing task. In this task,

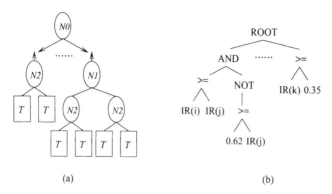

Fig. 3. The aspect of a typical controller. In this figure, *N0* is a dummy root node, *N1* represent logic components, and *N2* is the comparator $>=$. *T* can be a normalized sensor response or a threshold between 0 and 1 inclusive. The outputs of the subtrees are used to drive the actuators or activate another control system.

the robot is required to explore the given arena in order to find a box; once it detects the box, it then has to push the box toward a goal position indicated by a light source.

The task to be achieved is difficult for the following reasons. First of all, the robot is round, so that it only contacts the box at one point while pushing it, and the box tends to slide and rotate unpredictably when the pushing force exerted by the robot is not directed straight through the center of the box. Therefore, the robot has to adjust its own position occasionally in order to push the box forward as straight as possible. Furthermore, as there is no particular restriction on the initial relative positions of the robot, the box, and the ambient light, the robot can approach and detect the box at any position and orientation around the box; under such circumstance, the robot needs to move to a proper position deliberately in order to perform an efficient push to satisfy the final goal.

4.2 The Decomposition

To accomplish this task, we decompose it into two subtasks, *exploration* and *push-box-toward-light*. The former is to control the robot to explore the given arena in order to find the box without bumping into any wall; and the latter, to push the found box to a specific goal position. Again, the task *push-box-toward-light* is decomposed into two lower-level subtasks, *box-pushing* and *box-circling*. The goal of *box-pushing* is to keep the robot pushing a box forward, while the goal of *box-circling* is to keep the robot moving along the side of a box in order to provide the opportunity for the robot to move to suitable positions for pushing. Each of the subtasks, without being decomposed, is controlled by a separate behaviour primitive, and the different sub-controllers for the same task are merged by an arbitrator. Figure 4 shows the decomposition results and the aspect of the corresponding architecture for the target task. After the decomposition, the GP system is used to evolve both behaviour primitives and arbitrators.

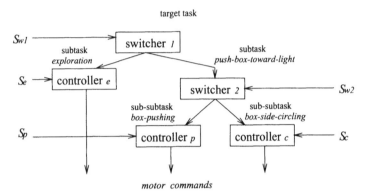

Fig. 4. The decomposition and integration of the target task. S_i indicates the sensory information relevant to control work i.

4.3   The Hardware Limitation

In the experiments below, we will use the method of evolving controllers in simulation and then testing them on a real Khepera robot. The simulator is built by employing a look-up table approach which has been shown to provide a close match between simulation and reality [12].

To accomplish the task *push-box-toward-light*, the arbitrator needs infra-red sensors and ambient light sensors to detect the box and the light respectively; and the ambient light sensors must be higher than the box to make sure they can detect the light even in the situation that the box is between the robot and the light. On the other hand, for the overall task, the other arbitrator will require certain perception information which can be organized as some sensory conditionals for the robot to recognize the box, to determine when to perform *exploration* and when to perform *push-box-toward-light*. For this purpose, we define a kind of virtual sensor *DR*, which can give the normalized reading difference between a pair of upper and lower infra-red sensors (The sensor pair here means two sensors pointing at the same direction but with different heights: one is higher and the other is lower than the box). A straightforward way to satisfy the requirements for both arbitrators is to duplicate the eight sensors of Khepera on its top (the sensors on the top are assumed to be higher than the box) so that the duplicated sensors can serve as ambient light sensors for the first arbitrator and as infra-red sensors for the second arbitrator (see Figure 5). However, the preliminary tests showed that when the robot was within a certain area around the bulb, the infra-red sensors on the Khepera robot were seriously disturbed by the normal bulb light and thus could not function properly. This would cause difficulty in verifying the simulation results on the real robot. Therefore, for the behaviour primitives which involve IRs only, we will evolve them in simulation and test them on the real robot; but the two arbitrators will be only evolved in simulation in which we assume that there are eight extra sensors on the top of the simulated robot as described above.

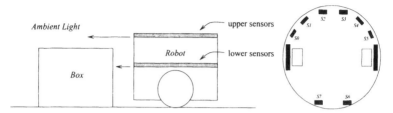

Fig. 5. (a) A possible way to satisfy the sensor requirements for both arbitrators is to duplicate a set of eight sensors on the top of the robot; (b) the sensor arrangement – a sensor S_i can function as an infra-red or an ambient light sensor.

5 Experiments and Results

5.1 Evolving Primitives for Task *Box Pushing*

As mentioned above, the task of *box-pushing* is that the robot should keep pushing a box forward as straight as possible. To achieve such a task, the robot needs to use its IR sensors to acquire perception cues for the location of the box. Therefore, we define two kinds of terminals, IRs and numerical thresholds, for our GP system to evolve controllers capable of achieving this task.

The fitness functions in this work are defined as penalty functions. For this task, the fitness function is formulated through the quantitative description of the expected behaviour, which is to keep the activation value of its front IR sensor high, the robot moving forward, and the speed difference of two motors low. The pressure from keeping the front IR sensor with high activation value is to reinforce the robot to head toward a box, and the pressure from keeping robot moving forward with low speed difference is to encourage the robot move straight and to prevent it from getting stuck in front of a box. The combination of these will lead to a pushing-forward behaviour. Thus, the fitness function for evolving a behaviour controller of box-pushing can be defined as:

$$f = \sum_{t=1}^{T} \alpha * (1 - s(t)) + \beta * (1 - v(t)) + \gamma * w(t)$$

in which $s(t)$ is the average of normalized sensor activations of the front sensors IR2 and IR3; $v(t)$ is the normalized forward speed; and $w(t)$ is the normalized speed difference of two motors at each time step t. During the evolution, each controller was evaluated in multiple trials and the average fitness value was used to measure the performance of this controller.

The typical box-pushing behaviour of the simulated robot, when performing the evolved controller, is illustrated in Figure 6(a). After evolution in simulation, this was transferred to a Khepera robot. Figure 6(b) shows the typical behaviour of the real robot. The figure for real robot was obtained by setting LEDs on the tops of the robot and box and using a video tracking system to record their trajectories [10]. This controller was tested on the real robot many times and each time it started from an arbitrary position and heading around the box. During the tests, the robot always generated consistent behaviour: it turned to face the box and then to approach and push it.

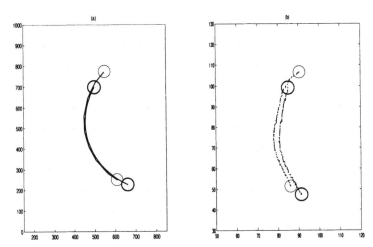

Fig. 6. The trajectories of simulated (left) and real (right) robots when they are pushing a box (the darker circles represent the boxes; and the boxes are pushed from top to down).

5.2   Evolving Primitives for Task *Box-Circling*

The task of *box-circling* is that the robot needs to keep moving forward and circling along the sides of a box. As in the box-pushing task, the robot should use its own IRs to capture the location of the box. Thus, terminals for evolving a controller to achieve this task are the same as those in the box-pushing task: IRs and numerical thresholds.

Again, we should define a fitness function to guide the evolution, and it can be done through the quantitative description of the expected behaviour: to keep the side sensor IR*0* with a certain activation value and the speed positive. The former is to encourage the robot to keep a certain heading relative to the box and a certain distance away from the box; and the latter is to reinforce the robot moving forward. The combination of these will produce a box-circling behaviour. Thus the fitness function is defined as:

$$f = \sum_{t=1}^{T} \alpha * abs(s(t) - k) + \beta * (1 - v(t))$$

where *abs* is a function which gives the absolute value of it argument; $s(t)$ is a normalized activation value of the specific sensor IRθ; k is a pre-defined constant indicating the distance between the robot and the box, in terms of the normalized sensor range; and v is the forward speed of a robot.

Figure 7(a) presents the evolved box-circling behaviour of the simulated robot, which shows that the task was achieved successfully in simulation. The evolved controller was then transferred to the real robot, and the typical behaviour of the real robot is demonstrated in Figure 7(b). We tested this evolved controller several times by putting the real robot around the box with an arbitrary heading each time. In all the tests, the robot showed similar behaviour: it performed turning to adjust its heading first and then moved along the side of the box. From the testing results, we can see that the robot is able to achieve the specified task reliably.

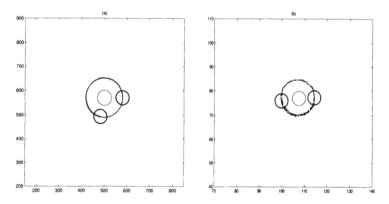

Fig. 7. The box-circling behaviours of simulated (left) and real (right) robots.

5.3 Evolving Primitives for Task *Exploration*

This task is that the robot needs to wander safely in an enclosure and visit as much of the enclosed space as possible. It can be described quantitatively as the space being divided into some squares and the robot must visit as many squares as possible during a fixed period of time. In the experiment below, we intend to evolve a reactive controller to explore the space without using the location information.

Unlike the experiments presented above, the fitness measurement for this task is not to sum up the penalty of each time step but to give a fitness value after a complete trial. The main concern for the fitness here is to minimize the number of squares which have not been visited, while an extra pressure on the speed is added to encourage the robot to move forward in exploring. Thus, the fitness function is defined as:

$$f = \alpha * (1 - P) + \beta * (1 - Avg)$$

in which P is the percentage of the space visited, i.e., $\frac{visited-squares}{total-squares}$; Avg is the average speed of the robot during a complete trial; and α, β are the corresponding weights. The enclosure in the exploration experiment here is a square of 50 × 50 cm^2 and each grid square is 5 × 5 cm^2.

As described above, the controller to be evolved is reactive and there is no location information provided here, so the robot must fully exploit its IR sensors to determine the turning angle carefully to achieve this task. Since IR sensors are the only mechanism for providing perception cues, the terminals for the exploration task are then defined to include IRs and numerical thresholds as in other tasks.

The exploration behaviour produced by the evolved controller is demonstrated in Figure 8(a), which shows that the robot is able to visit most of the specific arena during a fixed period of time. We should note that it is not important how the robot moves when it does not sense anything but the appropriate match of the turning angle (when the robot senses the wall) and the way it moves (when it does not sense anything) is nevertheless crucial for a reactive controller to perform exploration. As we can see in Figure 8(a), a successful match has been evolved and it enabled the robot to achieve the task.

Like the previous experiments, the evolved controller is downloaded to the real Khepera after the simulation. The behaviour observed from the real robot is presented in Figure 8(b). This behaviour is very similar to the one produced by the simulated robot in the simulated environment. Once again, it shows a successful example that we are able to evolve a behaviour controller by our GP system in simulation, and then transfer it to a real robot.

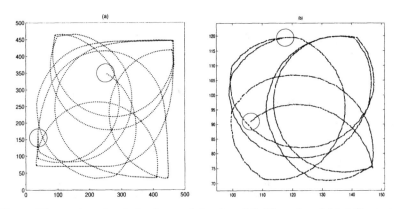

Fig. 8. The exploration behaviours of simulated (left) and real (right) robots.

5.4 Evolving Arbitrators for Task *push-box-toward-light*

As mentioned above, an arbitrator in this work is implemented as a reactive controller; its inputs are from the sensors and its outputs are used to trigger other controllers. For the arbitrator here, two kinds of sensors – infra-red and ambient light – are needed to detect the locations of the box and the light, so both kinds of sensors and numerical thresholds are defined as terminals to the GP system to constitute the structured sensory conditionals for the arbitrator. Since there are only two sub-controllers involved, the arbitrator is designated to have a single output to trigger them: if the output is 0, then the controller for subtask *box-pushing* dominates the control; otherwise the controller for subtask *box-circling* has the control. During the experiment, two subcontrollers will be frozen and only the arbitrator will be evolved.

In this task, the robot is expected to push the box as close as possible to the center of the area brightened by the light. Instead of measuring the distance between the goal position and the final position of the box at the end of a complete trial, we calculate the summation of the distance between the goal position and the box at each time step to reinforce the robot to push the box straight toward the light. Thus the fitness function is defined as:

$$f = \sum_{t=1}^{T} D_{b,l}(t)$$

in which $D_{b,l}(t)$ represents the distance between the box and the light source at each time step t.

Figure 9 illustrates, step by step, the typical behaviour of the simulated robot. As can be seen, the arbitrator first activated the primitive *box-circling* to move the robot along the side of a box. Once the robot reached an appropriate position in which the box was between the light and the robot itself, the control was then switched to the other primitive, *box-pushing*, to drive the robot to push the box forward. The *box-circling* and the *box-pushing* primitives were activated again in the same order if the box was not pushed directly toward the goal position. After the box was pushed to the goal position, the arbitrator continuously activated the primitive *box-circling* to circle the box in order to prevent pushing the box away from the box. From Figure 9, we can see that the box was successfully pushed to almost the center of the bright area.

We can also examine whether the performed task decomposition has been exploited in achieving the higher-level goal by observing the output sequence of the arbitrator: if the sequence is separated as periods of consistent activation, then the performed decomposition is confirmed to be helpful; otherwise it is not. Figure 10 demonstrates the output sequence corresponding to the behaviour shown in Figure 9. According to this figure, the evolved arbitrator is able to generate periods of quite consistent activation, except the short oscillating period – the transition period between two different behaviours – which did not effect the global behaviour at all. This figure, in fact, indicates that the involved lower-level sub-controllers have been fully exploited.

5.5   Evolving Arbitrators for the Overall Task

After evolving an arbitrator to combine two pre-evolved lower-level primitives, we can regard the integrated control system, including one arbitrator and two

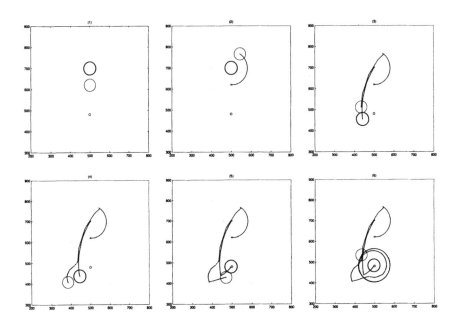

Fig. 9. The behaviour sequence of the robot: (1) The initial positions of the box (dark circle), the light (smallest circle) and the robot; (2) the robot moved along the side of the box; (3) pushing the box forward; (4) circling again to an appropriate position; (5) pushing the box again to the goal position; (6) continuously circling after the box has been pushed to the goal position.

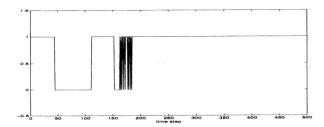

Fig. 10. The output sequence corresponding to the behaviour in Figure 9. In this figure, the y-axis indicates the controller which was activated: 0 is for *box-pushing* and 1 is for *box-circling*.

primitives, as a building block, and then evolve a new arbitrator to combine this building block and the other evolved controller *exploration* to achieve the overall task. As described in the above section, the arbitrator will need the perceptual information which can be used to recognize the appearance of the box, in order to generate proper output sequence to coordinate two involved control systems. Therefore, the virtual sensor *DRs* and numerical thresholds are defined as terminals for our GP system to evolve the desired arbitrator. As above, this arbitrator is designated to have one output: if the output is 0, the controller for *exploration* is activated; otherwise is the controller for *push-box-toward-light*. Again, the controllers to be combined will be frozen and only the arbitrator will be evolved.

The goal here is the same as the one above, to push the box as close as possible to the specified position, so the fitness function can be defined as above: to accumulate the distance between the box and the goal position at each time step. However, the criterion of simply measuring the fitness function for a fixed period time as before cannot give an objective evaluation here. This is because, in different trials, the robot could start from different positions and then take different numbers of time steps to find the box – it means that the lengths of the time of pushing are different. Therefore, in this experiment, the robot is given an extra period of time to find the box; and the fitness value is accumulated for a fixed period of time which starts from the moment the robot finds the box (or the end of the time period given for looking for the box). Thus, the fitness function is defined as:

$$f = \sum_{t=k+1}^{k+T} D_{b,l}(t)$$

in which $D_{b,l}(t)$ is the distance between the box and the goal position at time t; k is the time when the robot finds the box (or the end of the time period given for looking for the box); and T is the fixed period of time for fitness measurement.

The typical behaviour of the robot, when performing the whole control system, is shown in Figure 11. From these figures, we can see that the arbitrator firstly kept activating the controller *exploration* to drive the robot to explore the given environment and to avoid the walls. Once the robot found the box, the arbitrator began to activate the other controller, *push-box-toward-light*, according to the sensor stimuli. Since the arbitrator was able to activate this controller continuously after the robot found the box, the overall task was then achieved successfully. In fact, the performance of the arbitrator above can be observed from Figure 12, which shows the output sequence generated by this arbitrator during the test. It clearly indicates that this arbitrator could keep activating the controller *exploration* before the box was found, and afterward it can activate the other controller *push-box-to-light* consistently to achieve the target task.

6   Conclusion

In this work, we have analysed the difficulties of applying an evolutionary approach to learn complex behaviours for mobile robots. Instead of evolving the control system as a whole, we propose to take the control architecture of a behavior-based system and to learn the separate behaviours and arbitration by the use of an evolutionary approach. To assess the performance of the developed approach, we have evolved a control system to achieve an application task of box-pushing as an example. Experimental results show the promise and efficiency of the presented approach.

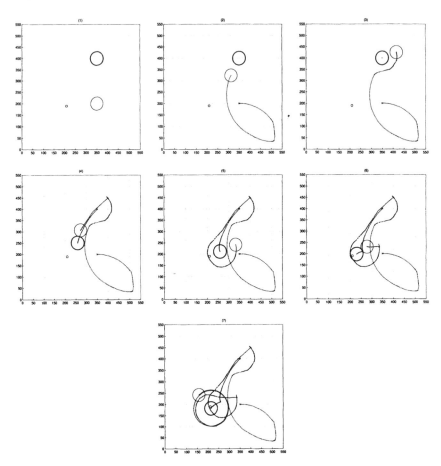

Fig. 11. The behaviour sequence of the robot: (1) the initial conditions; (2) the robot wandered around the environment before found the box; (3)∼(7) the robot continuously performed the building block controller *push-box-toward-light* to achieve the task.

The proposed approach has some advantages. First of all, by using the technique of task decomposition, the job of defining fitness function becomes more straightforward and the tasks become easier to achieve. In fact, for all the evolution experiments, the GP system converged to stable and sufficient solutions within only 30 generations. In addition to that, our controllers only involve logical operators, such as AND OR NOT, that are very simple to evaluate. This means that our approach is computationally cheap; and the evolved controllers can be easily compiled to custom hardware such as FPGAs to speed up the evaluation in controlling a robot.

Some further research work is currently in progress. For example, we have been constructing new sensors on the real robot to test the behavior arbitrator. We have also been investigating whether this approach can be applied to evolve control systems for even more complicated tasks.

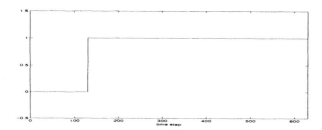

Fig. 12. The output sequence corresponding to the behaviour in Figure 11. In this figure, the y-axis indicates the controller which was activated: 0 is for *exploration* and 1 is for *push-box-toward-light.*

References

1. P. Agre, D. Chapman. Pengi: an Implementation of a Theory of Activity. In *Proceedings of AAAI-87*, pp. 268–272, Morgan Kaufmann, 1987.
2. R. A. Brooks. A Robust Layered Control System for a Mobile Robot. *IEEE Journal of Robots and Automation*, vol RA-2(1), pp.14-23, 1986.
3. D. Cliff, I. Harvey, and P. Husbands. Explorations in Evolutionary Robotics. *Adaptive Behavior*, 2(1):73-110, 1993.
4. M. Dorigo, M. Colombetti. Robot Shaping: Developing Autonomous Agents through Learning. *Artificial Intelligence*, 71(2):321-370, 1994.
5. R. J. Firby. Task Networks for Controlling Continuous Processes. In *Proceedings of the Second International Conference on AI Planning Systems*, pp.49-54, 1994.
6. D. Floreano and F. Mondada. Automatic Creation of an Autonomous Agent: Genetic Evolution of a Neural-Network Driven Robot. In *From Animals to Animats: Proceedings of the Third International Conference on Simulation of Adaptive Behavior*, pp.421-430. MIT Press/Bradford Books, 1994.
7. E. Gat. Robot Navigation by Conditional Sequencing. In *Proceedings of IEEE International Conference on Robitics and Automation*, pp.1293-1299, 1994.
8. W.-P. Lee, J. Hallam, H. H. Lund. A Hybrid GP/GA Approach for Co-evolving Controllers and Robot Bodies to Achieve Fitness-Specified Tasks. In *Proceedings of IEEE International Conference on Evolutionary Computation*, 1996.
9. W.-P. Lee, J. Hallam, H. H. Lund. Applying GP to Evolve Behaviour Primitives and Arbitrators for Mobile Robots. In *Proceedings of IEEE International Conference on Evolutionary Computation*, 1997.
10. H. H. Lund, E. V. Cuenca, J. Hallam. A Simple Real Time Mobile Robot Tracking System. Technical Paper No.41, Department of Artificial Intelligence, University of Edinburgh, 1996.
11. M. Mataric. Reward Functions for Accelerated Learning. In *Proceedings of International Conference on Machine Learning*, pp.181-189, 1994.
12. O. Miglino, H. H. Lund, S. Nolfi. Evolving Mobile Robots in Simulated and Real Environments. *Artificial Life*, 2(4), 1996.
13. S. Nolfi, and D. Parisi. Evolving Non-trivial Behaviors on Real Robots: An Autonomous Robot that Picks up Objects. In *Proceedings of the Fourth Congress of the Italian Association for Artificial Intelligence*. Spring-Verlag, 1995.

14. R. Pfeifer, and C. Scheier. Sensory-Motor Coordination: the metaphor and beyond. In *Robotics and Autonomous Systems, special issue in Practice and Future of Autonomous Agents*, 1996.
15. S. J. Rosenschein, L. P. Kaelbling. A Situated View of Representation and Control. *Artificial Intelligence*, vol 73, pp.149-174, 1995.
16. S. J. Rosenschein, L. P. Kaelbling. The Synthesis of Digital Machines with Provable Epistemic Properties. In *Proceedings of Conference on Theoretical Aspects of Reasoning about Knowledge*, pp.83-98. Morgan Kaufmann, 1986.
17. L. Steels. Building Agents out of Autonomous Behavior Systems. In L. Steels and R. Brooks (eds), *The Artificial Life Route to Artificial Intelligence*. Lawrence Erlbaum Associates, 1993.

Robot Learning Using Gate-Level Evolvable Hardware

Didier Keymeulen[1], Kenji Konaka[2],
Masaya Iwata[1], Yasuo Kuniyoshi[1],
Tetsuya Higuchi[1]

[1] Electrotechnical Laboratory
Tsukuba, Ibaraki 305 Japan
[2] Logic Design Corp.
Mito, Ibaraki 305 Japan

Abstract. Recently there has been a great interest in the design and study of evolvable and autonomous systems in order to control the behavior of physically embedded systems such as a mobile robot. This paper studies an evolutionary navigation system for a mobile robot using an evolvable hardware (EHW) approach. This approach is unique in that it combines learning and evolution, which was usually realized by software, with hardware. It can be regarded as an attempt to make hardware "softer". The task of the mobile robot is to reach a goal represented by a colored ball while avoiding obstacles during its motion. We show that our approach can evolve a set of rules to perform the task successfully. We also show that the evolvable hardware system learned *off-line* is robust and able to perform the desired behaviors in a more complex environment which is not seen in the learning stage.

1   Introduction

Robotics has until recently developed system able to automate mostly simple, repetitive and fixed tasks. These robots, e.g., arm manipulators, are mostly programmed in a very explicit way and in a well-defined environment. However for mobile autonomous robot applications, the environment is perceived via sensors and is less well defined. This implies that the mobile robot must be able to learn to deal with an unknown and possibly changing environment. Evolvable hardware [12] has been proposed as a novel approach to design real-time autonomous hardware systems. EHW refers to a class of hardware which modifies its hardware structure using evolutionary learning according to the rewards received from the environment [12].

In this paper we tackle the navigation task for a mobile robot which must reach a goal from any given position in an environment while avoiding obstacles. The robot is regarded as a *reactive system*, that is, without internal states, so that the robot behavior is based only on the current sensory inputs. The reactive system can be described by a Boolean function which is represented by a disjunctive normal form easily be implemented by a gate-level evolvable hardware.

Fig. 1. Real Robot

We show that the gate-level evolvable hardware is able to find a hardware configuration which implements the Boolean function, representing the tracking-avoiding reactive behavior of the robot, from a small set of observable sensor and motion pairs. We also show that EHW generalize well. This is necessary for the *off-line evolution*, where EHW is first learned with a "quasi-exact" simulation before used in the tracking-avoiding task, to maintain the robot performance in the real world despite the difference between the simulated and real world. Finally the hardware implementation is necessary for two main reasons. First it increases the computation speed of the controller needed for extremely fast real-time robot control. Second it speeds-up the adaptation of the desired behavior by simulating at the hardware level robot controllers for the *on-line evolution* approach, where the data gathered during tracking-avoiding task is used continuously to adapt EHW.

The paper first defines the robot task and its environment in section 2. In section 3, it describes the reactive navigation system based on a Boolean function controller represented in its disjunctive normal form. In section 4 we present the evolution mechanism used to find the Boolean function controller. In section 5 and section 6 we present the implementation of the Boolean function and the evolution mechanism on EHW. Finally in section 7 we illustrate the advantages of the EHW for off-line evolution.

2 Robot Environment and Task

The robot has a circular shape with a diameter of 25 cm (Fig.1). It has 10 infra-red sensors of 1 bit resolution situated at its periphery and equally distributed. To reduce the sensor space, the 10 infra-red sensors are mapped into 6 Boolean variables indicating the presence of objects at a distance less than 30 cm in 6

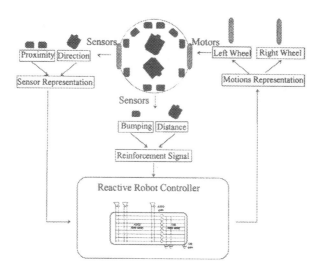

Fig. 2. Reactive Navigation System

directions. The robot is equipped with two cameras able to identify and track a colored object (the ball in Fig. 1). The vision module returns one of the 4 sectors, covering 90 degrees each, in which the target is located. It also returns the distance from the robot to the colored object, deduced from the tilt angle and the zoom of the camera tracking the object. The robot is driven independently by two motor wheels at both sides. They allow the robot to perform 8 motions: 2 translations, 2 rotations and 4 combinations of rotation and translation. The robot is controlled by a PC mother board connected to two transputer boards. One transputer board deals with the infra-red sensors, the vision module and the motor wheels. The other controls the EHW board executing the robot behavior. The environment is a world with low obstacles such that the colored target can always be detected by the robot. The obstacle shapes are such that using only a reactive system the robot will not be stuck. In other words, there are no overly complex shapes such as horseshoes. For the off-line evolution approach, we have built a "quasi exact" robot simulation to generate a robust behavior and evaluate the performance of the gate-level evolvable hardware.

The task assigned to the robot is to reach the target without hitting obstacles from any positions in the real world within a minimum number of motions and that . To perform its task, the robot must learn two basic behaviors, obstacle avoidance and going to the target, and coordinate these two behaviors to avoid becoming stuck due to repetition of an identical sensor-motor sequence.

3   Reactive Controller

May new ideas have been proposed to meet the challenges posed by mobile robots. Brooks decomposes a control system into a set of interacting behavior modules described in LISP-like programming languages [2]. Others use production rules approach as building blocks for the control system [6] [9] or general dynamical systems [25]. Some researchers propose to use artificial neural networks due to its generally smoother search space and its working with very low-primitives avoiding using preconceptions about the properties of the systems[11][13][1].

The use of complex building blocks for the control system of the robot is guided by the believe that all behaviors performed in the real world demands complex control structures. In our experiments we show that for the Markovian tracking-avoiding task, used by many researchers to show the validity of their approaches, a simple multi-outputs Boolean function control system is sufficient [18]. The multi-outputs Boolean function does not assume any knowledge of the necessary behaviors and of the high level primitives of each behavior. It is able to perform the navigation task in a reactive way and is well suited for an evolutionary search algorithm. It is also easily implemented in hardware. But for performing more complex tasks which are not Markovian such as a navigation task in an environment with obstacles of arbitrary shape or where the target is not always visible, exploiting complex control structures may be necessary.

The robot controller works as follows (Fig. 2). For each infra-red sensors, the controller extracts in a synchronous manner the presence of the obstacles and the direction of the target. It transforms this information into a Boolean vector. Based on this current Boolean vector, the controller chooses a motion to perform using the multi-outputs Boolean function. The output values of the function is the binary representation of the motion encoding the motor speed for the left and right wheel. In a synchronous way, the controller determines the speed of both wheels and then repeats the cycle of the operation.

Formally, the controller is a function $\mathcal{F}$ from the sensor input states to motions. The Boolean function approach describes the function $\mathcal{F}$ as m Boolean functions of n Boolean variables which represent the desired reactive behavior. The domain of input variables is $\{0,1\}^n$ where 2^n are the number of possible world states directly observable by the robot. It is encoded by 8 variables in our study: 6 for the infra-red sensors and 2 for the direction sensor. It represents 256 world states observable by the robot. The output range of the function $\mathcal{F}$ is $\{0,1\}^m$ where 2^m are the number of possible motions. In other words,

$$y = \mathcal{F}(x_0, \ldots, x_{n-1})$$

where $x_i \in \{0,1\}$ codes the proximity and direction sensors and $y \in \{0,1\}^m$ codes the 2^m possible motions.

A possible representation of function $\mathcal{F}$ is a look-up table [15]. But algorithms which represent the function $\mathcal{F}$ as a table present two disadvantages. First they are in general impractical for real-world applications due to their space and time requirements. Second these algorithms completely separate the information they have about one input situation without influencing what the robot will do in similar input situations. We choose to represent the m Boolean function f_i of the function $\mathcal{F}$ in terms of disjunction of purely conjunctive forms with a

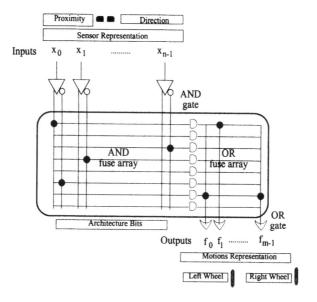

Fig. 3. Evolvable Hardware Controller (EHW).

limited number k of disjunctive terms such that they can easily be implemented in hardware.

$$\mathcal{F} \stackrel{\text{def}}{\equiv} \quad y = f_{m-1}(x_0, \ldots, x_{n-1})\, 2^{m-1} + \cdots + f_0(x_0, \ldots, x_{n-1})$$

where

$$f_u(x_0, \ldots, x_{n-1}) = (x_{i_u} \wedge \cdots \wedge \overline{x_{j_u}})_0 \vee \cdots \vee (x_{k_u} \wedge \cdots \wedge \overline{x_{l_u}})_{k-1}$$

and $u = 0, 1, \ldots, m-1$

4   Evolvable Reactive Controller

The goal is to identify the unknown function $\mathcal{F}$, mapping 256 possible world states (sensor states) to 8 outputs (motions), in a search space of 8^{256} functions, from a set of observable input-output pairs and a reward signal. From a machine learning perspective [21], the tracking-avoiding *control* task where the motions performed by the robot influence its future input situations [14] is an *associative delayed reinforcement* problem where the robot must learn the best motion for each world state from a *delayed* reward signal [26] [29][4].

Researchers have proposed for learning complex robot tasks, systems using previously learned knowledge such as explanation-based neural networks [27]. For simpler robot tasks in a less determined environment, researchers have proposed reinforcement learning algorithms [14]. Using this approach and to reduce the convergence time, Connell et al. integrates knowledge such as the properties of the task, the sensor configurations and the environment [4]. Asada et al. decomposes the input state space [20] [28]. For simple robot tasks in a unknown and dynamic environment, researchers have applied evolution-based learning algorithms to low level control architecture such as LISP-like programming languages [17][24][22], production rules (classifier systems) [30] [5] [10] and neural networks [3] [7] [23][1]. For our robot task and control architecture, we choose the evolutionary approach because we are most interested in navigation in a unknown environment.

The evolutionary approach uses genetic algorithm (GA) and operates on a population of bit strings each of them representing a function $\mathcal{F}$ and determining the behavior of a robot. The performance of all the behaviors for the tracking-avoiding task is evaluated and the resulting performance measures the fitness of the corresponding function $\mathcal{F}$. The GA gives better bit strings a higher chance to survive and reproduce. This way we get a new generation of bit strings which might incorporate good features of their parents. Each bit strings of this new generation is then used to represent the functions $\mathcal{F}$s and make behave the robots accordingly. Their performance is again evaluated and this process continues that way.

5 Evolvable Hardware

We implement the function $\mathcal{F}$ described by m Boolean functions by a gate-level EHW for two reasons. First to increase the computation speed of the controller for extremely fast real-time robot control. Second to increase the computation speed of the evolution mechanism to adapt the robot controller for *on-line* evolution.

In our experiment, the gate-level EHW consists of logic cells and a fuse array as shown on Fig. 3. The logic cells perform logical functions such as AND or OR. The fuse array determines the interconnection between the input and the logic cells. Each input includes its value and its inverse. If a column on a particular row of the fuse array is switched on, then the corresponding input signal is connected to the row and thus to the entrance of the logic cell. The fuse array is programmable. So in addition to the input bits and the output bits of the logic circuit simulating a Boolean function, the architecture of the EHW is determined by the architecture bits. Each link of the fuse array corresponds to a bit in the architecture bits [12].

The EHW structure suited for the implementation of function $\mathcal{F}$ consists of an AND-array and an OR-array (Fig.3). Each row in the AND-array calculates the conjunction of the inputs connected to it, and each column of the OR-array calculates the disjunction of its inputs. The number of columns of the AND array and the OR array is determined respectively by the number of inputs and outputs of the EHW. The number of rows of the arrays depends on the complexity of the disjunctive form of the m boolean functions connected to the output.

When the function $\mathcal{F}$ has one Boolean output f_0, in the worst case (f_0 is a xor of n boolean variables) the disjunctive form of f_0 needs 2^{n-1} purely conjunctive terms and its EHW implementation needs 2^{n-1} rows. When the function $\mathcal{F}$ has multiple Boolean outputs f_i we should have one AND-array for each Boolean output. But to reduce the search space, the m Boolean functions f_i share the same AND-array of the EHW in our architecture. In the worst situation (f_i is a xor and f_j is a $\overline{xor}$) the EHW for multiple Boolean outputs needs 2^n lines. In our experimental set up with a total of $n = 8$ Boolean variables, 6 for the infra-red sensors and 2 for the vision system and with $m = 3$ Boolean outputs, the EHW has $2*n+m = 19$ columns and needs a maximum number of $2^n = 256$ rows. However, the number of rows can be reduced by merging rows with the same output. In our experiments, we were able to reduce the number of rows to 50 and represent the EHW architecture with 950 architecture bits. In this way, on average , $\frac{256}{50} \simeq 5$ inputs states are mapped to the same output.

6 Evolvable Hardware Reactive Controller

For learning pure Boolean functions of the inputs from delayed reinforcement signal, the evolutionary method works in a similar way as Kaelbling's Generate-and-Test algorithm [14] but it uses a limited number terms. It is better suited for large search spaces when a large number of inputs and motions make $\mathcal{Q}$ learning impractical [29]. The evolutionary algorithm performs an incrementally parallel search in the space of Boolean functions, searching for a function $\mathcal{F}$ that exhibits high performance in the environment. It generates new Boolean functions by combining the existing functions using syntactic conjunction and disjunction operators. Statistics related to the performance of the functions in the environment are used to guide the search, choosing appropriate functions to be combined. Our evolutionary method uses genetically inspired search operators and selection mechanism.

The algorithm is implemented in hardware where the 950 architecture bits of EHW are regarded as the chromosome for the genetic algorithm[8]. If these bits change then the architecture and the function of hardware also change. The basic idea is to let the architecture bits evolve in order to adapt the function of the hardware to perform the tracking-avoiding task.

To analyze the transformation of function $\mathcal{F}$ by the genetic operators, we can interpret the EHW representation of function $\mathcal{F}$ in the following way (Fig.4). Each purely conjunctive term (EHW AND-array row) defines a hypercube of the input space where all inputs for which 1 is the correct output are on the cube and all inputs for which 0 is the correct output are outside the cube. The hypercube is called the support of the term. The support of Boolean functions f_i (EHW OR-array output) is the union of the hypercubes defined by each term of the disjunctive form (implemented by the rows connected to the EHW output). In the next section we analyze the mutation and cross-over operators.

6.1 Genetic Operators

Mutation Operator. There are five kinds of mutation operators: removing or setting a connection in the OR array, removing a connection on the AND array, setting directly and inversely a connection in the AND array.

The effect of mutation on the Boolean function is different if it is done on the architecture bits of the OR-array and the AND-array (Fig.4). In the OR-array, the mutation removes or adds one hypercube vertex to the support of one output of the OR-array. The change in the fitness depends on the overlapping of the hypercube vertex of the mutated line of the OR-array with the other hypercubes vertices that define the support of the EHW output of the OR-array. If it is completely overlapped there is no change in the fitness of the individual and the mutation will be *neutral*.

If the mutation is done on the architecture bits of the AND-array, it changes the shape of the hypercube defined by the mutated AND-array line. If the mutation removes/adds one connection then the size of the hypercube grows/decreases along the dimension associated with the added/removed variable. The third kind of mutation turns a direct into an inverse connection or the contrary. This mutation "translates" the support of the AND-array line along the dimension of the mutated connection. We chose as mutation rate is 0.05.

Different mutations have different effects on the lines. In our experiment, the mutation that enlarges the support of one line is rarer than the mutation which translates the support. The reason is to direct the evolution of the lines from small supports to larger ones. Once a line is fixed its support can grow.

Cross-over Operator. For a direct reinforcement problem, we are fairly tempted to try to evolve the lines and then build the function by selecting the best combination of lines. However for a delayed reinforcement problem, the robot can be stuck in a loop by executing a sequence of motions, e.g. going forward to reach the target and then backward to avoid the obstacle, and still be at the same world state perceived by the robot. To avoid that the learned behavior is stuck in a loop, we can not evaluate a line alone but a set of lines [10]. In order to evolve the best combination of lines for a Boolean function that needs severals lines, the technique we use is to cross-over the lines between the chromosomes (Fig.5). In that way the structures that remain unchanged between the generations are the lines. So the pressure of selection efficiently operate on the combination of lines. The selection keeps only the lines that perform well along with the other lines in the chromosome.

Genetic Parameters In our experiment we use a tournament selection with tournament size $s = 5$. In the tournament selection, s individuals are picked at random from the old population, and the fittest of them is put in the mating pool. This is repeated until the mating pool is as large as the old population. The tournament size s can be used to tune the selection pressure. The population size is 20. Elitism is used.

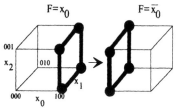

AND array mutation: inverse connection

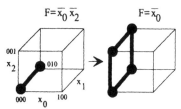

AND array mutation: delete connection

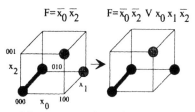

OR array mutation: add connection

AND-OR array cross-over

Fig. 4. Mutation of a Boolean function f_i of 3 variables using EHW: (1) Translation of the Support, (2) Decrease or Grow the Support, (3) Combine the Support

Fig. 5. Cross-over of two Boolean functions of 3 variables using EHW: Combine the Support

6.2 Fitness Evaluation

Each robot is evaluated in the simulated environment. It fails when it cannot reach the target within a maximum period of time. It can fail for two reasons:

– it hits an obstacle: this is detected by computing the intersection between the environment walls and the robot geometry. In the real world this is obtained from the bumping sensors.
– it reaches the maximum number of steps it is allowed to move in the environment. This situation occurs when the robot is stuck in a loop.

The fitness Φ of an individual is defined by a scalar between 0 (worst) and 1 (best) through combining two factors:

– the distance to the target $D(robot, target)$: it forces the robot to reach the target.

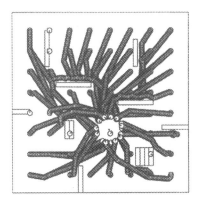

Fig. 6. Learning Off-Line Environment with Convex Obstacles

- the number of steps used to reach its actual position from its initial position. It forces the robot to chose a shorter path and to avoid to be stuck in loop.

In our experiment both factors have an equal contribution to the fitness:

$$\Phi = 0.5 \left(1 - \frac{D(robot,\ target)}{Environment\ Dimension} \right) + 0.5 \left(1 - \frac{Nbr.\ Steps}{Maximum\ Nbr.\ Steps} \right)$$

7 Experiments

In this section we present the experimental results of the evolvable hardware controller for off-line evolution. In the off-line approach both the EHW and the environment are first simulated to find the best EHW controller able to track a colored object and avoid obstacles. Second, the architecture bits of the best EHW controller is down loaded into the robot evolvable hardware board and control the robot in the real world. In the next section we present the method to build a robust EHW controller able to perform satisfactorily in a wide variety of environments.

In the off-line evolution approach, robustness is very important for two reasons. First there are so many combinations of sensory input and motions $(256 * 8 = 2048)$ that it is unlikely that the robot will ever experience most of them in the simulation. Second the difference between the simulated and the real environment is such that the robot performance may decrease in the real environment. To increase the robustness of navigation systems, some researchers either include noise in the simulated environment [13]. In our approach on one hand the generalization ability of EHW, obtained by its disjunctive normal form representation and the genetic operators, allows the robot to extrapolate what it has seen so far to similar situations which may arise in the future. On the other hand we shape the robot behavior by forcing the robot, during its evolution, to encounter many different situations.

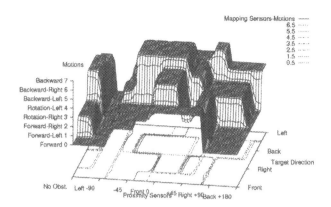

Fig. 7. Learning Off-Line Sensors-Motions Mapping of best Individual at generation 200

7.1 Simulated Environment with Convex Obstacles

We have conducted experiments with an environment containing 9 obstacles of different but all convex shapes. The distance between the obstacles is such that the robot can only perceive one obstacle at the same time, to limit the number of sensor inputs encountered in the environment. The target is situated in the middle right of the environment (Fig.6).

During evolution the individual is always placed at the same initial position in the environment: the upper left corner. At the beginning of evolution, all controllers are initialized at random. The probability to connect, respectively, directly or indirectly the inputs to the AND gates is both set to 0.3. The probability to connect the output of the AND gates to the OR gates is set to 0.5. The behavior of each individual is then simulated until it hits an obstacle, is stuck in a loop or when the robot reaches the target. When an individual reaches the target position, it is moved to a new position in the environment and its fitness is increased by 1. There are 64 new positions which are distributed equally in the environment and selected in a deterministic way introducing a shaping bias in the learning process. To keep a memory of the past success full experiences, all individuals always start from the same initial position and that the sequence of initial positions is always the same for all individuals.

The behavior of the best individual in the population at generation 200 is shown in Figure 6. It demonstrates that the robot coordinates obstacle avoidance and target tracking behaviors very well. For example it is able to turn right or left depending on the target direction when it encounters an obstacle. Figure 7 shows the sensor-motions mapping for the 48 representative input sensor states: 12 obstacle positions for 4 possible target directions. It shows the ability of EHW to generalize. For example in Fig. 7 the robot executes a forward motion when

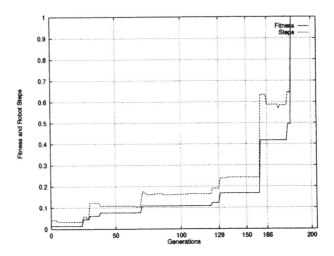

Fig. 8. Fitness (1 = 64 new positions) and number of robot steps (1 = 3015 motions) of the best individual throughout generations.

there is an obstacle at its left side independently of the target direction.

Figure 8 shows the number of steps and the fitness of the best individual with generations. First we observe that the number of training steps (nbr. of generations ∗ nbr. of individuals ∗ nbr. of motions) is a factor of 10 to 100 less than the number of training steps used by evolution algorithms applied to neural networks [19] or to production rules [5]. Second we observe two types of jumps during the evolution due respectively to mutation and cross over operators. The mutation operator creates individuals avoiding obstacles and generates a large jump of the fitness with a small jump of the number of steps. For example at generation 129 after a mutation, the best individual is able to avoid an obstacle, to reach the 10th new position and then it hits another obstacle after 575 steps. The cross over operator creates individuals going in less motions to the target and generates a small jump of the fitness with a large jump of the number of steps. For example at generation 166 after a cross-over, the best individual is able to shorten its path to target from 1911 to 1770 by finding a better combination of sensor-motion pairs but it is still hitting the same obstacle.

7.2 Simulated Environment with Concave Obstacles

We have also conducted experiments with an environment containing 5 convex and 4 concave obstacles. Figure 9 shows the behavior of the best individual in the population at generation 285 in an environment with concave obstacles during one experiment. It demonstrates that the evolution discovers that turning clockwise around the obstacle until there is no obstacle in the direction of the target works well. The behavior of the best individual is robust enough to be applied

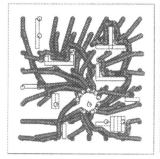 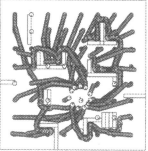 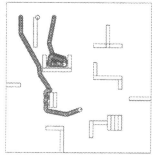

Fig. 9. Learning in Environments with concave obstacles.

Fig. 10. Testing the robust behavior in a more complex environment with concave obstacles.

Fig. 11. Hand Coded solution for the complex environment with concave obstacles.

in a different and even more complex environment as shown in the behavior of the robot around the right bottom obstacle in Figure 10. A hand coded strategy using 11 lines mutually exclusive is unable to pass around the concave obstacles as shown in Figure 11

7.3 Real Environment with Concave Obstacles

Once the robot reaches the target 64 times, we download the controller of the best individual at generation 285 into the EHW board. To evaluate the performance of the controller in a real environment where the initial condition, the target position and the shape and the position of the obstacles are different from those in the simulated environment.

The real world differs from the simulated world on at least three aspects. First the motor in the real world are controlled by continuous movement while in the simulation they are controlled by discrete steps. Second the model of the sensors behavior and the motor execution is not the same as the real world. Third the positions, shapes and orientations of obstacles are different. However, the evolved EHW controller is capable of reaching the target in a unseen environment. Video of the robot behavior can be seen at *http://www.etl.go.jp/etl/kikou/ didier*.

The off-line approach supposes that the simulated world is designed carefully and that the EHW controller is robust enough to deal with the uncertainty of the sensor values and motors. Unfortunately this approach cannot take into account failures of the robot hardware itself and does not allow the controller to be self-adaptive, e.g., changing its structure while performing its task. One approach to make the robot adaptive and to reduce the number of interactions of the robot with the environment is to build an *on-line* evolutionary system. The robot is allowed to do the evolution process in an internal model of the environment learned during the robot's interaction with the environment and the evolution process can be accelerated by the utilization of evolvable hardware [16].

8 Conclusion

In this paper we have presented an evolutionary navigation system for mobile robots. The navigation task is to track a colored target in the environment while avoiding obstacles. The mobile robot is equipped with infra-red sensors and a vision system furnishing the direction and the distance to the colored target. The navigation control system is completely reactive and is based on a gate-level EHW implementing a Boolean function in its disjunctive form.

We demonstrate the feasibility to implement an evolvable Boolean controller using EHW. We have shown that a complete reactive navigation system is able to perform the tracking and avoiding task. We have also shown and demonstrated the generalization ability of the EHW controller. The generalization ability allows on one hand to build off-line a highly robust control system which is insensitive to the shape of the obstacles and the position of the target and on the other hand to maintain the robot performance in the real and simulated environment despite the difference between the two environments.

The EHW implementation has been designed to work as a stand-alone navigation system in a mobile robot to accelerate the learning process and to open the door for *on-line* EHW adaptation to real world applications [16].

Acknowledgments

This research was supported by MITI Real World Computing Project (RWCP). The authors would like to express great thanks to Dr. Otsu and Dr. Ohmaki of ETL for their support and encouragement, Prof. Yao from University of New South Wales for his invaluable comments and to Prof. Hoshino of University of University for his valuable discussions.

References

1. Randall D. Beer and J.C. Gallagher. Evolving dynamic neural networks for adaptive behavior. *Adaptive Behavior*, 1(1):91–122, July 1992.
2. Rodney Brooks. Artificial life and real robots. In F. J. Varela and P. Bourgine, editors, *Proceedings of the First European Conference on Artificial Life*, pages 3–10, Cambridge, MA, 1992. MIT Press / Bradford Books.
3. Dave Cliff, Inman Harvey, and Philip Husbands. Explorations in evolutionary robotics. *Adaptive Behavior*, 2(1):73–110, July 1993.
4. Jonalthan H. Connell and Sridhar Mahadevan. *Robot Learning*. Kluwer International Series in Engineering and Computer Science. Kluwer Academic Publisher, 1993.
5. Marco Dorigo and Marco Colombetti. Robot shaping: developing autonomous agents through learning. *Artificial Intelligence*, 71:321–370, 1994.
6. Marco Dorigo and Uwe Schnepf. Genetics-based machine learning and behavior-based robotics: A new synthesis. *IEEE Transactions on Systems, Man, and Cybernetics*, 23(1):141–154, 1993.

7. Dario Floreano and Francesco Mondada. Automatic creation of an autonomous agent: Genetic evolution of a neural-network driven robot. In J-A. Meyer Dave Cliff, Philip Husbands and S. Wilson, editors, *From Animals to Animats 3: Proceedings of the 3rd International Conference on Simulation of Adaptive Behavior*. MIT Press, 1994.

8. David E. Goldberg. *Genetic Algorithms in search, optimization and machine learning*. Addison Wesley, 1989.

9. John J. Grefenstette. Incremental learning of control strategies with genetic algorithms. In *Proceedings of the Sixth International Workshop on Machine Learning*, pages 340–344. Morgan Kaufmann, 1989.

10. John J. Grefenstette and A. Schultz. An evolutionary approach to learning in robots. In *Proceedings of the Machine Learning Workshop on Robot Learning, Eleventh International Conference on Machine Learning*. New Brunswick, NJ, 1994.

11. Inman Harvey, Philip Husbands, and Dave Cliff. Seeing the light: Artificial evolution, real vision. In J-A. Meyer Dave Cliff, Philip Husbands and S. Wilson, editors, *From Animals to Animats 3: Proceedings of the 3rd International Conference on Simulation of Adaptive Behavior*. MIT Press, 1994.

12. Tetsuya Higuchi, Tatsuya Niwa, Toshio Tanaka, Hitoshi Iba, Hugo de Garis, and T. Furuya. Evolvable hardware with genetic learning: A first step towards building a darwin machine. In Jean-Arcady Meyer, Herbert L. Roitblat, and Stewart W. Wilson, editors, *Proceedings of the 2nd International Conference on the Simulation of Adaptive Behavior*, pages 417–424. MIT Press, 1992.

13. Philip Husbands, Inman Harvey, Dave Cliff, and Geoffrey Miller. The use of genetic algorithms for the development of sensorimotor control systems. In F. Moran, A. Moreno, J.J. Merelo, and P. Chacon, editors, *Proceedings of the third European Conference on Artificial Life*, pages 110–121, Granada, Spain, 1995. Springer.

14. Leslie Pack Kaelbling. *Learning in Embedded Systems*. Bradford Book, MIT Press, Cambridge, 1993.

15. Didier Keymeulen, Marc Durantez, Kenji Konaka, Yasuo Kuniyoshi, and Tetsuya Higuchi. An evolutionary robot navigation system using a gate-level evolvable hardware. In *Proceeding of the First International Conference on Evolvable Systems: from Biology to Hardware*, pages 195–210. Springer Verlag, 1996.

16. Didier Keymeulen, Masaya Iwata, Kenji Konaka, Ryouhei Suzuki, Yasuo Kuniyoshi, and Tetsuya Higuchi. Off-line model-free and on-line model-based evolution for tracking navigation using evolvable hardware. In *Proceeding of the First European Workshop on Evolutionary Robotics*. Springer Verlag, 1998.

17. John Koza. Evolution of subsumption using genetic programming. In F. J. Varela and P. Bourgine, editors, *Proceedings of the First European Conference on Artificial Life*, pages 3–10, Cambridge, MA, 1992. MIT Press / Bradford Books.

18. Henrik Hautop Lund and John Hallam. Evolving sufficient robot controllers. In *Proceedings of the International Conference on Evolutionary Computation*, pages 495–499, Piscataway, NJ, 1997. IEEE.

19. Orazio Miglino, Henrik Hautop Lund, and Stefano Nolfi. Evolving mobile robots in simulated and real environments. *Artificial Life*, 2(4):417–434, summer 1995.

20. T. Nakamura and M. Asada. Stereo sketch: Stereo vision-based target reaching behavior acquisition with occlusion detection and avoidance. In *Proceedings of IEEE International Conference on Robotics and Automation*, pages 1314–1319. IEEE Press, 1996.

21. Balas K. Natarajan. *Machine Learning: a theoretical approach*. Morgan Kaufmann Publisher, 1991.
22. Peter Nordin and Wolfgang Banzhaf. An on-line method to evolve behavior and to control a miniature robot in real time with genetic programming. *Adaptive Behavior*, 5(2):107–140, 1997.
23. Domenico Parisi, Stefano Nolfi, and F. Cecconi. Learning, behavior and evolution. In *Proceedings of the First European Conference on Artificial Life*, pages 207–216, Cambridge, MA, 1992. MIT Press / Bradford Books.
24. Craig W. Reynolds. An evolved, vision-based model of obstacle avoidance behavior. In *Artificial Life III*, Sciences of Complexity, Proc. Vol. XVII, pages 327–346. Addison-Wesley, 1994.
25. Luc Steels and Rodney Brooks, editors. *The Artificial Life Route to Artificial Intelligence: Building Embodied, Situated Agents*. Lawrence Erlbaum Assoc, 1995.
26. Richard S. Sutton. Special issue on reinforcement learning. *Machine Learning*, 8(3/4), 1992.
27. Sebastian Thrun. *Explanation-based Neural Network Learning: A Lifelong Learning Approach*. Kluwer Academic Publishers, Boston, MA, 1996.
28. Eiji Uchibe, Minoru Asada, and Koh Hosoda. Behavior coordination for a mobile robot using modular reinforcement learning. In *Proceedings of IEEE/RSJ International Conference on Intelligent Robotits and Systems (IROS 96)*, pages 1329–1336. IEEE Press, 1996.
29. Christopher J.C.H. Watkins and Peter Dayan. Q-learning. *Machine Learning*, 8(3):279–292, 1992.
30. Stewart Wilson. Classifier systems and the animat problem. *Machine Learning*, 2:199–228, 1987.

Lecture Notes in Artificial Intelligence (LNAI)

Lecture Notes in Computer Science